D1711764

Outsourcing War to Machines

Outsourcing War to Machines

The Military Robotics Revolution

PAUL J. SPRINGER

Praeger Security International

 PRAEGER™

An Imprint of ABC-CLIO, LLC

Santa Barbara, California • Denver, Colorado

Library of Congress Cataloging-in-Publication Data
Names: Springer, Paul J., author.
Title: Outsourcing war to machines : the military robotics revolution / Paul J. Springer.
Description: Santa Barbara : Praeger Security International, [2018] | Includes bibliographical references and index.
Identifiers: LCCN 2017044702 (print) | LCCN 2017050745 (ebook) | ISBN 9781440830860 (ebook) | ISBN 9781440830853 (alk. paper)
Subjects: LCSH: Military robots. | Military robots—Moral and ethical aspects. | Military art and science—Technological innovations— Moral and ethical aspects.
Classification: LCC UG479 (ebook) | LCC UG479 .S67 2018 (print) | DDC 355.8—dc23
LC record available at https://lccn.loc.gov/2017044702

ISBN: 978-1-4408-3085-3
EISBN: 978-1-4408-3086-0

21 20 19 18 17   1 2 3 4 5

This book is also available as an eBook.

Praeger
An Imprint of ABC-CLIO, LLC

ABC-CLIO, LLC
130 Cremona Drive, P.O. Box 1911
Santa Barbara, California 93116–1911
www.abc-clio.com

This book is printed on acid-free paper ∞

Manufactured in the United States of America

Contents

Preface   vii

1   The Revolution Has Arrived   1

2   A Short Guide to Revolutions   23

3   The Long Gray Line   57

4   Automating the Battlefield   79

5   Robot Lawyers   115

6   Morality for Machines?   147

7   The Global Competition   169

8   The Road Map   195

Notes   221

Bibliography   239

Index   253

A photo essay follows page 114.

Preface

I am a military historian by training and trade, who commenced his career examining the American treatment of enemy prisoners of war. This topic is very Army-centric and no doubt contributed to my opportunity to teach history at the U.S. Military Academy at West Point. I loved my time at the U.S. Military Academy and worked with outstanding colleagues and wonderful cadets—but there was a darker side to the position. I taught at West Point from 2006 to 2009, during the worst of the fighting in Iraq, and the vast majority of my cadets were deployed to a war zone within a year of graduating from the academy. Too many of those former cadets did not come home, a fact that began to weigh very heavily upon me and no doubt drove me to accept a position at the Air Command and Staff College. This book emerged, in large part, out of a piece of advice I received from my dissertation advisor, Professor Brian Linn of Texas A&M University. When I told him that I was accepting the new position, he opined, "You had better start examining some topics that the Air Force cares about." As usual, Brian was 100 percent correct. I began to examine some of the emerging aspects of 21st-century warfare and publishing about them in parallel with my forays into 19th-century U.S. Army history.

Military robots are enabling an entirely new form of warfare, one that will fundamentally alter the dynamics of human conflict. To their proponents, they allow bloodless war, an opportunity to punish the evildoers of the world without placing one's own forces at risk. To their opponents, they are eroding the natural barriers against the commencement of warfare, which naturally includes the risk to one's own forces. It is my contention that both sides of this argument are correct—military robots do allow the conduct of warfare with impunity, from a distance. However, the very fact that an armed attack becomes so easy makes an armed attack infinitely more likely. Thus, they serve to reduce some (but certainly not all) of the consequences of a war and make the decision to go to war carry far fewer political costs. When the United States decided to engage in a series of airstrikes against Libya in 1986 (Operation El Dorado Canyon), President Ronald Reagan and his closest advisors spent weeks discussing the potential ramifications before deciding to conduct a single night of aerial bombardment against very specific targets. They knew full well that they were risking the lives of the pilots involved and that they might be condemned by the international community for what was undoubtedly an act of war. The last three American presidents (George W. Bush, Barack Obama, and Donald Trump) have all had at their disposal the ability to attack against individual targets anywhere in the world through remotely piloted aircraft. Collectively, they have ordered airstrikes in at least a dozen countries, none of them an active conflict zone, in pursuit of members of al Qaeda and the Islamic State, including U.S citizens. It is hard to say how much each president wrestled with the decision to launch strikes into the sovereign territory of other nations—but it is undeniable that the means to do so eroded any prohibitions against the activity.

Military robots are here to stay—there is no way to put the genie back in the bottle, and they are simply too effective as a tool of war to be voluntarily relinquished. However, to date, no nation has fielded a fully autonomous platform allowed to make lethal decisions without human intervention in anything but a defensive role. The capability to build such platforms already exists, and, when they are unleashed upon the battlefield, they will likely to be extremely effective, so much so that other nations will scramble to produce their own variants. Coupled with the increased willingness to use force as a tool of diplomacy, such weapons are likely to trigger a catastrophic conflict that will quickly spread around the globe, unless preventive measures are taken to avert this development. This work seeks to place the development of such machines into an historical context and then examine the likely futures if the current path does not change.

As any author can attest, no book is ever completed in isolation. I have been fortunate to have the support of family, friends, and colleagues in this endeavor. In particular, in 2015, I was fortunate to be selected as the chair of the Department of Research at Air Command and Staff College. Accepting the position created ridiculous delays in the production of this manuscript, which tried the patience of my editor, Pat Carlin, to its very limits. However, the position has been the most rewarding period of a very happy career, and I owe the fellow members of the department a very deep vote of gratitude. Professor Kenneth Johnson has served as the deputy chair for my entire stint at the head of the department, and the two of us are of one mind on virtually every major decision. Lieutenant Colonel Steven Quillman has been the voice of reason and restraint, and his fellow plank-holder, Lieutenant Colonel Paul Clemans, has always found ways to improve the processes of our organization. Our more recent additions, Lieutenant Colonel Peter Garretson and Major Brent Ziarnick, have proven to be a formidable team that has managed to rewrite the entire school curriculum regarding space power. Dr. Lisa Beckenbaugh and Dr. Jonathan Zartman have each taken a significant aspect of the Air Command and Staff College experience and made it more student friendly and at the same time more challenging for our graduates. It is to this group of colleagues and scholars, who I hold in the highest esteem and cherish both individually and collectively, that I dedicate this work.

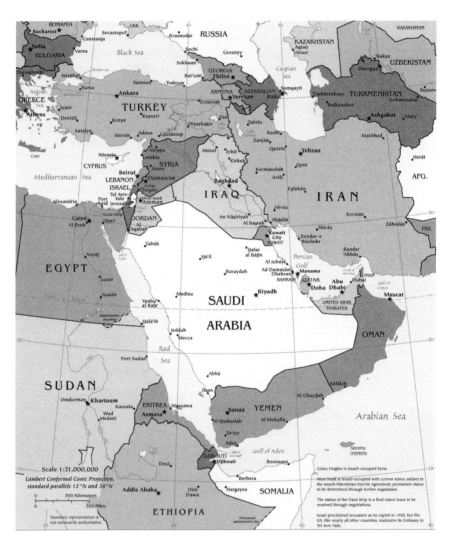

Political map of the Middle East (*CIA World Factbook*)

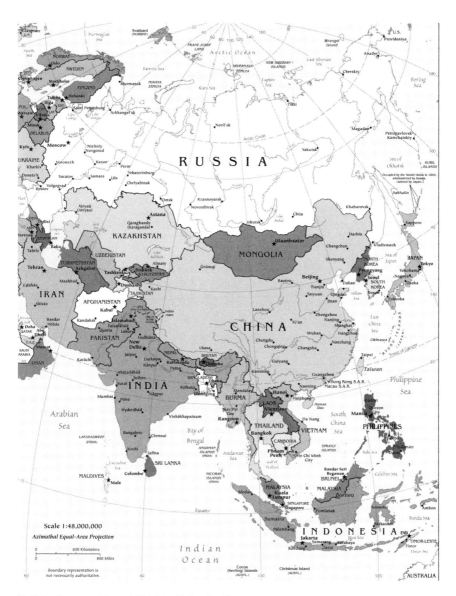

Political map of Asia (*CIA World Factbook*)

CHAPTER 1

The Revolution Has Arrived

Be not deceived. Revolutions do not go backward.

—Abraham Lincoln, 1856

On July 24, 2017, a terrorist bomber detonated a motorbike-driven rickshaw packed with explosives in a busy market in Lahore, Pakistan, killing 26 people. Pakistani police forces were the primary target of the bombing and comprised 9 of the dead, but a further 17 bystanders were killed in the blast. The incident was well reported around the world, with international leaders expressing sorrow at the attack.[1] Shortly after the incident, Tehrik-i-Taliban Pakistan (TTP) took credit for the attack, which they considered another blow in their insurgency against the Pakistani government. In particular, the TTP has been frustrated by the Pakistani government's willingness to ignore U.S. airstrikes in Pakistani territory, including one that killed the founder of the group, Baitalluh Mehsud, in 2009.[2] Mehsud, who had quickly become the most dangerous man in Pakistan, was hunted by armed drones for months before being struck by a missile while lounging on a rooftop in South Waziristan, surrounded by his family. In previous attempts to kill Mehsud, armed drones had fired dozens of missiles, killing at least 100 people, including 70 in a single strike on a funeral for one of Mehsud's lieutenants.[3] When Mehsud died, President Barack Obama crowed about the achievement in a live radio broadcast, happily acknowledging that the United States had launched the missile strike.[4]

Four months earlier, a suicide bomber infiltrated an ammunition storage depot in Balakleya, Ukraine, and detonated a thermite grenade in a stack of ammunition crates. The resulting chain reaction of explosions required the evacuation of 20,000 citizens and devastated a significant portion of the city. The attack destroyed military ordnance worth approximately US$1 billion, with untold property damages in the surrounding area. Yet, it received almost no press coverage, perhaps in part because of the low number of casualties—to date, a single death has been attributed to the attack. No group has formally taken responsibility for the attack—but it demonstrated a new form of attack that has been used to devastating effect in the past two years: suicide drones. Small, remotely guided flying robots have been able to sneak past defenses designed to protect against aerial attacks, ground assaults, and human infiltration to deliver their small but deadly packages. Thermite is a mixture of aluminum powder and iron oxide, which burns at approximately 4,000 degrees Fahrenheit when ignited. This allows a single canister, weighing less than 2 pounds, to cut through a half-inch thick slab of steel—and to burn anything contained inside. On at least four occasions, Ukrainian ammunition dumps have been attacked by the same method, with devastating results.

The attacks were almost certainly carried out by Russian operatives, given the state of conflict between Russia and Ukraine. What made the attacks particularly noteworthy, beyond the sheer scale of the destruction they caused, was the underlying method being used. In each case, the explosives were carried into the targeted facilities by a small, commercially available unmanned aircraft. A quad-copter, similar to the hottest holiday gift of 2016, was slightly modified so that its remote operator could place its deadly cargo and trigger it (an action that destroyed the drone in each case). The aircrafts' operators needed only to fly over the fencing of the compounds and locate a stack of ammunition crates, an action taking only a few seconds, to inflict catastrophic damage upon the Ukrainian military, far beyond what has been witnessed in the Predator and Reaper strikes usually associated with attacks by drones.[5] The attacks demonstrated that even small payloads, delivered to the right location, can have enormous strategic effects, a form of attack that David Hambling has dubbed "bringing the detonator" and relying upon the local environment for the bulk of the destruction.[6]

A NEW FORM OF CONFLICT FOR A NEW CENTURY

For almost the entire twenty-first century, the United States has been involved in an international conflict with terrorist organizations, particularly al Qaeda and the Islamic State, but also affiliated (and nonaffiliated) groups sprinkled around the globe. For most young American citizens, particularly those approaching military age, conflict has become the new norm—the United States has literally been at war for as long as they can remember. Because the war is as much a fight against ideology and specific activities as it is a struggle with enemy organizations, it presents special challenges to civil and military leaders charged with bringing it to a successful conclusion. Many have turned to advanced technology to create an asymmetrical advantage, in the hopes that using high-performance machines might offset the challenges of attacking a system of beliefs that have diffused throughout the world.

Military robots have become a part of international conflict in the twenty-first century. While the technology is still evolving, there is little chance that it will disappear, because it has simply proven too useful to be abandoned. As Mark Bowden notes,

The drone is effective. Its extraordinary precision makes it an advance in humanitarian warfare. In theory, when used with principled restraint, it is the perfect counterterrorism weapon. It targets indiscriminate killers with exquisite discrimination. But because its aim can never be perfect, can only be as good as the intelligence that guides it, sometimes it kills the wrong people—and even when it doesn't, its cold efficiency is literally inhuman.[7]

Of course, there are downsides to the use of advanced technology. One unnamed intelligence official put it quite succinctly, stating,

"But there are two big ways we can make mistakes," he added. "One is to forget that sometimes a light footprint can cost you more in the long run than going into a place with a much more decisive force—that was the lesson of Afghanistan. And the second is to fall in love with a whiz-bang new technology, because it's easy to justify relying on it more and more. And that's when a tactical weapon can begin defining your strategy."[8]

Thus far, there have been enormous successes in the War on Terror, and there have been abject failures. The use of military robotics has contributed to both.[9]

PURPOSES AND AVOIDANCES

This book seeks to provide context to the rise and deployment of military robotics. It raises issues with the legality and morality of using these advanced systems and critiques the ways in which they have been used in recent conflicts. This book is not an attempt to reverse the path of the development of military robotics—such an outcome is almost impossible to imagine and is well beyond the capabilities of the author. However, it points out that on a number of occasions, short-term political gains were sought at the cost of magnifying long-term dangers, and at times there seems to have been almost no thought given to the precedents set by the United States in its War on Terror. As James Sullins notes, "It is probably impossible to contemplate the alternative anymore, but we should have avoided arming robots in the first place."[10] What has been done cannot be reversed—but the worst potential consequences might still be avoided with the application of significant forethought and strategic planning.

This work is primarily centered around the behavior of the U.S. government in its current conflicts. This is in large part because the United States is driving much of the technological change—and has become the earliest and most prolific user of military robotic systems in the twenty-first century. Although there are discussions of other nations and their behaviors in this new realm of conflict, most nations are watching the American experience before determining their own paths. Thus, the precedents set by the United States in its War on Terror are likely to have a much longer and deeper effect than many U.S. decision makers seem to recognize. Or, as Richard Falk puts things, "The embrace of state terror to fight against non-state actors makes war into a species of terror and tends toward making limits on force seem arbitrary, if not absurd."[11] To Falk, the primary weapon of the War on Terror is not the weaponized drone, per se, but the willingness to fly such weapons over enormous swaths of territory around the world as a means to keep millions of foreign citizens permanently cowed.

In the pursuit of context, Chapter 2 discusses revolutions in military affairs (RMA), which are essentially fundamental transformations of the modes and methods of human conflict. It puts forth the argument that military robotics, particularly those with autonomous power to make lethal decisions, are a transformative technology that will permanently alter human warfare if they are built and deployed in significant numbers. Chapter 3 provides an overview of the development of robotics, from the earliest attempts at remote-controlled weaponry to the decision in 2001 to attach missiles to a Predator aircraft previously dedicated to unarmed intelligence, surveillance, and reconnaissance missions. Chapter 4 discusses the decision to place military robots into active service in the War on Terror and outlines the major milestones in the twenty-first century that included such platforms. It is not a comprehensive examination of every type of military robot currently in service (such a discussion could fill volumes) but rather a discussion of the most prominent models that have had the most significant effects thus far. Chapter 5 analyzes the development of the laws of armed conflict and how the utilization of military robotic systems might require changes to them. It also examines how the availability of military robots is altering American governmental behavior, especially with regard to the use of force and the protections from government violence previously considered inalienable for U.S. citizens. Chapter 6 discusses the ethics of warfare, including their development and categorization and how robots might be influencing the morality of conflict. It uses a number of short case studies to illustrate the key principles of morality and then demonstrates how robots might either enhance or destroy the concept of a just war. Chapter 7 examines a number of conflict zones throughout the world where military robotics are having definite effects on the behavior of governments and where the boundaries of human-machine conflict might be pushed ever further than by the events of the War on

Terror. Chapter 8 discusses a series of potential future scenarios involving military robotics and is as close as this volume gets to any attempts at prognostication.

DEFINITIONS

A few key terms require specific definitions, which will be germane to the remainder of this work. Although these definitions are not perfect, they are functional—they do not markedly differ from the definitions used by other authors, engineers, or government entities. That said, these definitions certainly do not match the common usages in English for the same concepts, particularly by laypersons. Journalists, especially, have a pronounced (and unfortunate) habit of using terms that are convenient, if imprecise, and of using words with radically different implications as if they were interchangeable. While the current work will seek to avoid such errors, they will be unavoidable within direct quotations from time to time. The reader is encouraged to fix the following concepts firmly in mind, if only to avoid falling into the trap of clouded thinking that so many armchair strategists are already in when discussing the current and future aspects of military robotics.

A "robot" is a machine that is capable of sensing its environment, at least in some fashion, and choosing its behavior based upon that sensory input. That behavior must include some form of action in the physical world; thus, a computer program cannot in and of itself be considered a robot.[12] That does not imply that a robot has an unlimited number of options or that it needs to perceive the world in the same fashion as its human creators—such a perception is currently impossible and would be inadvisable under almost any circumstances. Just as replicating human sensory inputs renders an enormous engineering problem, it also creates a set of unnecessary restrictions. Humans, for example, cannot perceive in the infrared or ultraviolet portions of the light spectrum, but machines can be built with such capabilities. Humans struggle to differentiate aromas, particularly when they are blended, whereas machines can be imbued with the ability to separate and analyze multiple odors or chemical signatures. The key considerations for using the term "robot" are the sensory functions and the autonomous selection of behavior from multiple options. Examples of common robots include industrial manufacturing systems, which might be required to change their behaviors depending upon the task at hand or to stop performing a task if the safety of a human operator might be placed at risk.

A "robotic system" is a machine that incorporates some, but not all, of the conditions necessary for use of the term "robot." Most often, this term is accurately used to refer to remotely piloted vehicles (RPVs), which may have limited autonomous functions. For example, the RQ-4 Global Hawk is an enormous, unmanned reconnaissance aircraft. It is not remotely flown in the sense of more familiar RPVs, as it performs the vast majority of its flight activities without human input. However, it does not decide for itself whether or not to engage in a mission, and it flies only where it is ordered to go. In order to remain aloft, the Global Hawk must sense its environment, which includes changes in altitude, airspeed, weather conditions, and the presence of other aircraft in its vicinity. However, it also has a sensor suite that is subject to human controls—in essence, the Global Hawk airframe primarily exists as a platform to deliver its sensory equipment to the airspace above or near a reconnaissance target, and the Global Hawk does not rely upon that sensory data to perform its own functions.

A drone is a machine that follows a set of instructions without any degree of autonomy—it simply does what it has been programmed to do, regardless of environmental factors. This is by far the most misused of military robotic terms in the twenty-first century, and it is primarily inappropriately applied to remotely piloted aircraft being used for airstrikes in the War on Terror. The term "drone " more accurately refers to target platforms that fly preprogrammed routes for gunnery practice or reconnaissance aircraft that simply fly a planned route without a human pilot in control, even from a distance. The term most certainly should not be used to refer to combat platforms like the MQ-1 Predator or its larger cousin, the MQ-9 Reaper, both of which are under the positive control of a human operator and which cannot utilize any lethal weapons without a deliberate act by the operator. Calling such a machine a drone grammatically absolves the operator of any actions that might result from its utilization, as if lethal airstrikes simply "happened." A cruise missile might be considered a drone, although as Timothy Sundvall notes, "The line between cruise missile and UAV is beginning to be blurred by technology."[13] Richard M. Clark argues that cruise missiles do not rise to the level of the term "drone," in large part because they are only used a single time, making them only a very advanced projectile.[14]

A machine's "autonomy" refers to the level of decision making that it may undertake without human intervention. Kenzo Nonami, Farid Kendoul, Satorshi Suzuki, Wei Wang, and Daisuke Nakazawa collaborated to create a classification system for autonomous machines, in which they broke down the distinct levels of autonomy into 10 classes:

1. Remotely guided: certainly achieved by the Pioneer in 1986, arguably by earlier systems

2. Real-time health/diagnosis: the RQ-1 Predator began reporting system self-checks to human operators in 1995

3. Adapt to failures and flight conditions: the RQ-4 Global Hawk corrects its flight path based upon environmental conditions, without human input, and has done so since 1998

4. Onboard route replan: experimental systems are now rerouting themselves upon encountering obstacles or unexpected conditions

5. Group coordination: rudimentary swarming behavior

6. Group tactical replan: with or without a more advanced "controlling" system

7. Group tactical goals: determining tactical behavior to obtain operational objectives

8. Distributed control: cloud-based computing systems and advanced swarm behavior

9. Group strategic goals: providing desired political outcomes and allowing machines to pursue them

10. Fully autonomous swarms: complete self-controlled systems

These classifications are not ironclad—and many systems in development seem to fit into more than one category. However, by creating this hierarchy of semiautonomy, leading to the fully autonomous final stage, these researchers provide a very handy way to conceptualize the increasing sophistication of military robotics.[15]

The development of limited autonomy in machines has already occurred, but no completely autonomous machines have yet been designed, much less built. Linguists might parse the notion of limited autonomy—if one's freedom is curtailed, can one truly be considered free? Their objections aside, the concept of limited autonomy is a useful notion in the real world—and one that applies to humans as much as it might to robots. There are inherent limits upon human behavior, some of them created by external constraints and some of them due to internal mechanisms. External constraints are things beyond the control of the actor—and might include such issues as fiscal realities, physical limitations, or legal ramifications of certain actions. No matter how much I might desire to quit my profession and spend my entire life being pampered on a private yacht sailing around the world, I do not have the tens of millions of dollars that would be required for such a lifestyle. This imposes an external fiscal limit upon my ability to engage in a certain action. Perhaps I could steal the money necessary to fund such a life, although doing so would almost certainly result in law enforcement moves to halt my activities and to make certain that I never had the opportunity to set foot upon a yacht for years, if not the rest of my life. All of these constraints prohibit a certain activity that I might otherwise entertain. Internal limits also prevent me from such an act—even if I had the capability to steal millions of dollars and get away with the crime, my moral compass will not allow me to steal. My work ethic will not allow me to simply waste my life, and the resources that would be involved, roaming the oceans of the world, contributing nothing to society. These internal mechanisms govern my behavior just as surely as any external limits, even though they are theoretically under my control.

Robots are also governed by external and internal limits. The external factors include issues such as the construction of the robot itself as well as its placement and function. For example, the RQ-1 Predator aircraft is a relatively lightweight airframe. This means it can fly using a far lighter engine and hence requires less fuel for each mile traveled, but it also means that the aircraft does not function well in certain weather environments and has an extremely limited payload. If the designers wanted a larger payload, they had the ability to redesign the airframe for a more rugged utility, even if it remains essentially the same size. The redesigned version, the RQ-9 Reaper, can carry a much larger amount of hardware and fly slightly faster, but it does so at the cost of a heavier engine and hence a shorter range. Robots can use only the tools that they are given—an industrial paint robot in an automobile factory will never be expected to fly over hostile airspace and fire missiles at terrorists—just as a Predator is unlikely to put a metallic gloss coat on a new Hyundai. Unlike humans' physical form, created through thousands of generations of evolution, these robots' physical limitations are deliberately designed and are anything but accidental. Thus, the external limits placed upon robots are known at the outset and not subject to substantial changes without a significant amount of effort on the part of the designers.

Robots' internal limits, on the other hand, are largely derived from the programming instructions provided and can at least theoretically be modified relatively easily, in the form of software updates. Any computer user knows how quickly a programming update can change the function of a machine—and at the same time knows how vulnerable programming code can be to errors. Microsoft Corporation, makers of the most ubiquitous software in use on the planet, is forced to continually release updates of its products to end users. Often, these updates are designed to either fix how the program functions in relation to other software or close loopholes in the code by which the users' computer might be vulnerable to cyberattack. Users who fail to update their computer software on a regular basis become increasingly vulnerable to malicious code that can attack and compromise their computers.[16]

Currently, armed military robots can be classified into three primary categories, commonly shorthanded as "man-in-the-loop," "man-on-the-loop," and "man-out-of-the-loop." The first category applies to robotic systems that are controlled by a human operator and in particular that cannot engage in lethal behavior, such as firing a weapon, without a positive command from the human operator. The most prominent examples are the Predator and Reaper aircraft, which are under the positive control of a human pilot located in a remote control station. Man-on-the-loop systems have a substantial amount of autonomous function—for example, the RQ-4 Global Hawk autonomously handles most of the tasks associated with takeoffs, flight, and landings, but a human operator is capable of assuming control if necessary. Man-out-of-the-loop systems are allowed to handle their own responsibilities without human permissions or intervention—the most well-known example for American audiences is likely the Patriot missile battery, which can be placed in full autonomy mode. In that mode, the battery is capable of tracking a target and opening fire, subject to predefined parameters, without seeking human permission. In general, the currently deployed systems with this level of autonomy are in air defense and point defense situations, where human reaction times are simply too slow to allow effective utilization of the weapon.

A "cyborg" is a cybernetic organism. Essentially, it is an organic life form that has integrated some aspect of electronic technology to enhance its own capabilities. The term tends to evoke visions of science fiction films, yet it is a term that can be accurately applied to an increasing number of humans. In many ways, cyborgs have become so commonplace as to remain unnoticed, as their cybernetic implants tend to be a means of compensating for an organic failure or weakness, rather than any attempt to create inhuman powers. When Dr. William DeVries implanted the world's first artificial heart, on December 2, 1982, he turned Barney B. Clark into a cyborg. Without the artificial heart, Clark would have died due to acute cardiac failure coupled with the lack of a donor match. Although the implant extended his life only by 112 days, Clark lived those remaining days as a cyborg. One of the most prominent cyborgs in Western society is radio broadcaster Rush Limbaugh. In 2001, he was diagnosed with auto-immune inner-ear disease, a condition that would render him completely deaf and thus unable to continue his hugely successful broadcast career. Instead, doctors performed a cochlear implant, compensating for his body's inability to hear and allowing him to remain a prominent figure in talk radio. Pleased with the results, he underwent a second procedure to restore hearing in his other ear, in 2014.[17]

An individual utilizing a prosthetic limb is not considered a cyborg, unless that limb is directly tied into the individual's remaining organic tissue and responds to nerve impulses. Such prosthetics were unimaginable even 30 years ago but are now being used to restore function to wounded military veterans on a regular basis and are beginning to penetrate the civilian market. The public has become both more aware of amputees and more accepting of the use of prosthetics. One need only witness the Wounded Warrior Amputee Softball Team or the Wounded Warrior Amputee Football Team compete (and win) against able-bodied teams to understand the amazing progress that has occurred.[18]

ARTIFICIAL INTELLIGENCE AND THE NATURE OF COGNITION

What is the nature of thought, and how might it apply to the pursuit of artificial intelligence (AI)? Human thought is one of the most challenging concepts of modern science—and defining it has been the province of philosophers, mathematicians, physicians, biologists, and computer scientists for centuries. While we have learned a great deal about the biochemical processes associated with the human brain, we still do not have a unanimously accepted model of how thinking works in humans, much less a meaningful way to replicate it in machines. Perhaps our greatest mistake has been the attempt to duplicate organic processes in an inorganic environment. Machines designed to focus upon a particular aspect of thinking, such as arithmetic calculations or pattern recognition, have proven far more capable than humans in these hyper-specialized functions. Thus far, the most useful "thinking" machines have been those dedicated to very specific tasks, at which computers can excel. Because computers do not get tired of repetitive tasks, bored with inactivity, or too exhausted to concentrate, they offer unique possibilities with regard to functions that have traditionally remained human responsibilities. However, turning many military operations over to machines carries its own risk, as the conclusions drawn by computers regarding the same input available to humans may differ markedly from those of humans. Further, because computers have a much wider variety of potential sensory inputs, and because they are typically much better at reconciling conflicting information from varied sensors, they may act far more decisively than their human counterparts, even in situations when decisive action may not be warranted.

Early modern philosophers contemplating the nature of cognition developed a wide variety of explanations for human thought and how individual ideas might be stored within the mind. Thomas Hobbes considered "thinking" to be a form of mental discourse—and he made no differentiation between the process of thinking and the act of writing. To Hobbes, the most rational form of thinking followed methodical rules, which led to predictable patterns and reasoned conclusions. Rene Descartes divorced thought, which he considered inherently symbolic, from the objective world. In this regard, he essentially created the modern concept of the mind. Descartes believed the mind was entirely separate from the physical world, and thus reason should be considered entirely divorced from physical laws. David Hume attempted to discern the laws of the mind, applying scientific processes to the notion of mental mechanics. To Hume, thoughts were akin to movements of matter within the brain—an idea that offered the possibility of sharing thoughts directly if the said matter could be isolated and transferred to another individual. Each of these classical philosophers not only expanded the human conception of intelligence but also offered guidance for the development of artificial forms of thinking.

For pioneers of AI, it was the Hobbesian notion of methodical patterns that drove the concept of programming. Descartes's conception of symbolism rather than literalism within the mind advanced the AI concept of internal programming languages that need not conform to the rules of written language—and in fact, if they are to maintain an ironclad Hobbesian methodical approach and should not incorporate any of the idiosyncrasies of human languages. Hume's conception of thoughts literally moving through the mind became the concept of data transfers—and also the possibility of copying data for replication in other machine environments.

In the twentieth century, scientists began to actively pursue the concept of AI. In the early 1960s, computer and robotics pioneer Marvin Minsky founded the Artificial Intelligence Laboratory at the Massachusetts Institute of Technology. His former colleague and close friend John McCarthy founded the Stanford Artificial Intelligence Laboratory, developing a new key locus for AI research on the West Coast. Both sites remain groundbreaking robotic research locations that have made enormous contributions to the development of AI. Hundreds of other academic institutions, private corporations, and government entities have also moved into the field of robotics development in the ensuing five decades.

Predictions regarding the future of AI have ranged from the absurdly optimistic to the hopelessly pessimistic. In 1989, Paul Lehner proclaimed, "My first prediction is that most of the national defense applications of AI presently being pursued will not succeed in the near-term development of operationally useful systems, despite the fact that many of the programs have the specific objective of developing operationally functional systems in the near future."[19] Two years later, the AI systems controlling Patriot missile batteries proved quite successful in military operations, though not as infallible as some media reports suggested at the time. In 1993, Vernor Vinge opined, "[W]ithin thirty years, we will have the technological means to create superhuman intelligence. Shortly after, the human era will be ended."[20] One of the most prominent, and optimistic, futurists is Ray Kurzweil, who in 2003 claimed that "it is hard to think of any problem that a superintelligence could not either solve or help us solve."[21] Of course, if the problem proves to be a hostile artificial superintelligence, harnessing that capability might prove difficult. Even Kurzweil believes that controlling such a development might prove impossible, or, as he states the matter, "Once strong AI is achieved, it will immediately become a runaway phenomenon of rapidly escalating superintelligence."[22] He considers this to be a positive development in human history but seems unwilling or unable to consider the potential negative ramifications of such an achievement.

None of these predictions has proven correct, but then again, none of them was completely wrong, either. Lehner's prediction that there would be no operationally useful systems was obviously quickly proved incorrect—and yet, there are few operational autonomous systems nearly three decades after his prediction. Vinge's idea of superintelligence was wrong, if one requires that the AI in question be capable of every form of cognition practiced by humans. If one allows for limited applications, though, there are many aspects of information processing that are far more effective using computers. It is harder to assess Kurzweil's predictions, in large part because they have yet to occur, but they are unlikely to be hyper-accurate as well. By definition, predictions of the future are fickle (and if Greek mythology is correct, getting the right answer won't guarantee anybody will listen). This book does not pretend to have the answers, but it does help to address the questions of how we got to where we are in terms of military robotics, where we are likely headed, and what might occur in the future.

There is an old joke in the historical profession, which is an apt metaphor for the majority of this book. Like most historians' jokes, it follows a tried and true pattern:

Q: "How many historians does it take to change a light bulb?"

A: "When Thomas Edison invented the incandescent light bulb in 1879. . ."

This book, like most works produced by historians, seeks to place the events it describes into historical context. In part, this is a function of the training of the discipline. In part, it is because too many other works examining military robotics have utterly failed in this regard, leading the reader to assume that military robots simply appeared in the skies over Afghanistan one day, raining down missiles upon terrorists' heads. The truth, as usual, is quite a bit more nuanced and will require more examination.

THE ANTECEDENTS OF THE WAR ON TERROR

When al Qaeda operatives hijacked four airliners and turned them into suicidal missiles, a large share of the U.S. population had never heard of the terror organization. The U.S. State Department had placed al Qaeda on its list of foreign terror organizations in 1999, and the CIA stepped up efforts to find and neutralize its reclusive leader.[23] The Department of Defense had stepped up protective measures for troops operating in the Middle East due to recent overseas attacks. Average Americans, on the other hand, paid the group little or no mind, despite a series of earlier attacks on U.S. citizens and American interests. Ultimately, it was this conflict that created the conditions necessary to rush military robotics to the battlefield, with little consideration for the long-term consequences of such a decision. However, as is so often the case, there is a much more complicated backstory to the conflict with al Qaeda.

In 1993, a truck bomb detonated in the underground garage of the World Trade Center complex and came disturbingly close to penetrating the "bathtub" of the structure. Such a penetration would have allowed the waters of the Hudson River to rush in, an event that could only end in catastrophe for the buildings and their inhabitants. Contemporary investigators posited that such a breach might have resulted in the deaths of up to 50,000 people. Unlike the September 11 attacks, there would have been no opportunity for an evacuation, as all the routes of egress would be submerged almost immediately, and the buildings' collapse would have followed shortly thereafter. The mastermind of the 1993 attack, Ramzi Yousef, was captured in 1995 in Pakistan and is currently serving a life sentence in the supermax prison located at Florence, Colorado, after being convicted of plotting both the World Trade Center attack and the Bojinka plot of 1995. His Pakistani uncle, Khalid Sheikh Mohammed, took note of the failing aspects of the 1993 and 1995 plans and within a few years commenced the operational planning for the 2001 attacks that destroyed the complex.

Although the FBI was quickly able to determine responsibility for the attack, it had no way of knowing the scale of the rising threat presented by al Qaeda, a shadowy terror group formed by mujahedeen veterans from the Afghan resistance to Soviet occupation. The group's founder, Osama bin Laden, was a wealthy scion from one of the largest construction companies in Saudi Arabia. He offered the services of his organization to the Saudi government in August 1990, as an Iraqi invasion of Kuwait put hostile troops on the Saudi border. The Saudi leadership rebuffed bin Laden and chose to request assistance from the United States, inviting American-led coalition troops to establish bases along the border. The Desert Storm campaign of 1991 that ousted the Iraqis from Kuwait was launched from those bases and conducted without assistance from bin Laden's nascent organization. The choice of Western rather than local assistance infuriated bin Laden, who swore to rid the Muslim Holy Land of corrupting Western influence.

Over time, al Qaeda's goals morphed and grew, to the point that by 2001, the organization was determined to ignite a regional insurgency against all of the temporal governments of the Arab world. Ultimately, bin Laden hoped to bring about a pan-regional caliphate, uniting the world's Sunni Muslims into a single empire. If realized, such a state would stretch from Morocco to Indonesia, encompassing a billion adherents. It would be a world power impossible to ignore, with economic and population resources to challenge any other nation. To rule such a diverse polity, bin Laden expected to establish a theocracy ordained by Allah, which would blend political and religious leadership under a single figure. Naturally, sharia law would reign, in accordance with bin Laden's own Wahhabist views of Sunni Islam.

A savvy student of history and politics, bin Laden understood that fear of a common enemy has served as a key unifying force for the major empires of the world. To create such a common enemy, he needed to not only attack the West but also provoke it into a massive overreaction. Such a response might serve to galvanize and unite the disparate peoples of the

Sunni faith, particularly if their own governments were seen to side with the Western aggressors. In 1996, he served a formal notice of his intent to declare war upon the West and proclaimed it the duty of every faithful Muslim to resist the corrupting influence of Western ideology. In his declaration, bin Laden directly addressed U.S. secretary of defense William J. Perry and claimed justification for any acts of terrorism in the service of the greater good:

I say to you William (Defense Secretary) that: These youths love death as you love life. They inherit dignity, pride, courage, generosity, truthfulness and sacrifice from father to father. They are most delivering and steadfast at war. They inherit these values from their ancestors (even from the time of the Jaheliyyah, before Islam). These values were approved and completed by the arriving Islam as stated by the messenger of Allah (Allah's Blessings and Salutations may be on him): "I have been send to perfecting the good values" (Saheeh Al-Jame' As-Sagheer).

Those youths know that their rewards in fighting you, the USA, is double than their rewards in fighting some one else not from the people of the book. They have no intention except to enter paradise by killing you. An infidel, and enemy of God like you, cannot be in the same hell with his righteous executioner.

Terrorising you, while you are carrying arms on our land, is a legitimate and morally demanded duty. It is a legitimate right well known to all humans and other creatures. Your example and our example is like a snake which entered into a house of a man and got killed by him. The coward is the one who lets you walk, while carrying arms, freely on his land and provides you with peace and security.

The youths hold you responsible for all of the killings and evictions of the Muslims and the violation of the sanctities, carried out by your Zionist brothers in Lebanon; you openly supplied them with arms and finance. More than 600,000 Iraqi children have died due to lack of food and medicine and as a result of the unjustifiable aggression (sanction) imposed on Iraq and its nation. The children of Iraq are our children. You, the USA, together with the Saudi regime are responsible for the shedding of the blood of these innocent children. Due to all of that, whatever treaty you have with our country is now null and void.[24]

Although the CIA was aware of bin Laden's existence, analysts did not appear to feel very threatened by his intentions. A 1996 CIA profile of bin Laden characterized him as a financier of terrorism rather than as a charismatic mastermind representing a threat to the entire West.[25]

Two years later, bin Laden joined in a communiqué signed by the World Islamic Front declaring a "Jihad against Jews and Crusaders." Once again, the prevailing theme was opposition to Western dominance in Middle Eastern affairs. It is noteworthy that the declaration included Jews, and by extension the nation of Israel, which al Qaeda had previously largely ignored. This was almost certainly done to coopt the allegiance of Palestinian resistance organizations, who had access to thousands of trained fighters, advanced weaponry, and substantial financial reserves. Under this new declaration of jihad, the authors laid out a series of grievances and then called for all able-bodied Muslims to participate in the conflict to the extent of their abilities.

No one argues today about three facts that are known to everyone; we will list them, in order to remind everyone:

First, for over seven years the United States has been occupying the lands of Islam in the holiest of places, the Arabian Peninsula, plundering its riches, dictating to its rulers, humiliating its people, terrorizing its neighbors, and turning its bases in the Peninsula into a spearhead through which to fight the neighboring Muslim peoples.

If some people have in the past argued about the fact of the occupation, all the people of the Peninsula have now acknowledged it. The best proof of this is the Americans' continuing aggression against the Iraqi people using the Peninsula as a staging post, even though all its rulers are against their territories being used to that end, but they are helpless.

Second, despite the great devastation inflicted on the Iraqi people by the crusader-Zionist alliance, and despite the huge number of those killed, which has exceeded 1 million . . . despite all this, the Americans are once against trying to repeat the horrific massacres, as though they are not content with the protracted blockade imposed after the ferocious war or the fragmentation and devastation.

So here they come to annihilate what is left of this people and to humiliate their Muslim neighbors. Third, if the Americans' aims behind these wars are religious and economic, the aim is also to serve the Jews' petty state and divert attention from its occupation of Jerusalem and murder of Muslims there. The best proof of this is their eagerness to destroy Iraq, the strongest neighboring Arab state, and their endeavor to fragment all the states of the region such as Iraq, Saudi Arabia, Egypt, and Sudan into paper statelets and through their disunion and weakness to guarantee Israel's survival and the continuation of the brutal crusade occupation of the Peninsula.

All these crimes and sins committed by the Americans are a clear declaration of war on God, his messenger, and Muslims. . . . On that basis, and in compliance with God's order, we issue the following fatwa to all Muslims:

The ruling to kill the Americans and their allies—civilians and military—is an individual duty for every Muslim who can do it in any country in which it is possible to do it, in order to liberate the al-Aqsa Mosque and the holy mosque [Mecca] from their grip, and in order for their armies to move out of all the lands of Islam, defeated and unable to threaten any Muslim. This is in accordance with the words of Almighty God, "and fight the pagans all together as they fight you all together," and "fight them until there is no more tumult or oppression, and there prevail justice and faith in God."

We—with God's help—call on every Muslim who believes in God and wishes to be rewarded to comply with God's order to kill the Americans and plunder their money wherever and whenever they find it. We also call on Muslim ulema, leaders, youths, and soldiers to launch the raid on Satan's U.S. troops and the devil's supporters allying with them, and to displace those who are behind them so that they may learn a lesson.[26]

Also in 1998, al Qaeda operatives detonated two truck bombs at U.S. embassies in Nairobi, Kenya, and Dar es-Salaam, Tanzania. Over 200 victims perished in the attacks, the vast majority of them local African citizens working at or near the embassies. Less than two years later, a small boat loaded with explosives detonated next to the USS *Cole* while the American destroyer was at anchor in Aden, Yemen. Seventeen U.S. sailors died in the attack, which narrowly failed to sink the ship. Each of these attacks provoked a U.S. retaliation in the form of cruise missile strikes against al Qaeda training camps, but neither did much to influence the U.S. citizenry regarding the threat of terrorism in general or al Qaeda in particular. American military interest remained fixated upon the Balkans, in particular over the question of Serbian incursions into Kosovo. U.S. Air Force units continued to enforce no-fly zones over northern and southern Iraq. Little effort was expended upon countering the rise of al Qaeda or training forces in the tenets of counterinsurgency and counterterrorism operations.

When President George W. Bush entered office in 2001, he did so under the cloud of a contested election that had to be decided in the U.S. Supreme Court. His administration focused little attention upon al Qaeda or other terror organizations, and efforts by President Clinton's top counterterrorism expert, Richard Clarke, had little effect upon the new national security team. On August 6, 2001, the president's daily brief from the CIA focused upon the threat presented by bin Laden and his organization. In particular, the later-declassified brief noted that al Qaeda showed an unusually high capacity for complicated operations and long-term planning. It also summarized the previous mass-casualty attacks carried out by al Qaeda and suggested that bin Laden had taken a particularly strong interest in the possibility of hijacking commercial aircraft.[27] Bin Laden's desire to attack the United States in a spectacular fashion might have been self-evident to the intelligence and law enforcement agencies, as well as to the military, but the manner of such an attack remained unforeseen until it was too late.

It took most Americans completely by surprise to watch in real time as fires raged at the World Trade Center in New York. Hundreds of millions of people around the world watched in horror as first one tower and then the second collapsed into a mountain of rubble, entombing nearly 3,000 victims in the process. By the evening of September 11, it was clear that the United States had been the victim of the deadliest terror attack in history and that a new form of conflict had begun. President George W. Bush addressed the nation in the aftermath of the attack, stating:

Today our fellow citizens, our way of life, our very freedom came under attack in a series of deliberate and deadly terrorist acts. The victims were in airplanes or in their offices: secretaries, business men and women, military and Federal workers, moms and dads, friends and neighbors. Thousands of lives were suddenly ended by evil, despicable acts of terror . . .

Today our Nation saw evil, the very worst of human nature. And we responded with the best of America, with the daring of our rescue workers, with the caring for strangers who came to give blood and help in any way they could . . .

The search is underway for those who are behind these evil acts. I've directed the full resources of our intelligence and law enforcement communities to find those responsible and to bring them to justice. We will make no distinction between the terrorists who committed these acts and those who harbor them.[28]

This initial response was designed to calm the nation and demonstrate resolve in the face of terror. It is interesting that it made no mention of activating military forces or launching international campaigns, beyond the naked threat against any nation that might harbor al Qaeda.

If bin Laden thought the United States would pull back from the Middle East in response to the September 11 attacks, he was sorely mistaken. On the other hand, if his goal was to chip away at the economic foundations of U.S. power, the attacks succeeded beyond his wildest dreams. In exchange for an investment of approximately US$500,000, al Qaeda triggered American involvement in two wars, the creation of an enormous security apparatus, and a social division within American society that questioned whether fundamental citizens' rights were being sacrificed in the name of pursuing an unobtainable goal of perfect security. The costs of all of the subsequent U.S. military activities and security upgrades have probably run to at least US$5 trillion and will likely continue to drive the U.S. government deeper into debt.[29]

In the days that immediately followed the attacks, Bush and his closest advisors were largely devoted to formulating a military response and to designing a campaign to destroy Osama bin Laden's terror network, headquartered in Afghanistan under the protection of the Taliban government. Bob Woodward reports that one of the most significant aspects of the response was the massive increase in lethal authorizations for the CIA, noting,

Late in the afternoon at the White House, the president was presented with two documents to sign. One was a Memorandum of Notification modifying the finding that President Ronald Reagan had signed on May 12, 1986. The memorandum authorized all the steps proposed by Tenet at Camp David. The CIA was now empowered to disrupt the al Qaeda network and other global terrorist networks on a worldwide scale, using lethal covert action to keep the role of the United States hidden. The finding also authorized the CIA to operate freely and fully in Afghanistan with its own paramilitary teams, case officers, and the newly armed Predator drone.[30]

During a September 20 speech to Congress, the tenor of Bush's remarks became far more belligerent and left no doubt that the United States was girding itself for war:

On September 11th, enemies of freedom committed an act of war against our country. Americans have known wars, but for the past 136 years, they have been wars on foreign soil, except for one Sunday in 1941. Americans have known the casualties of war, but not at the center of a great city on a peaceful morning. Americans have known surprise attacks but never before on thousands of civilians. All of this was brought upon us in a single day, and night fell on a different world, a world where freedom itself is under attack.

Americans have many questions tonight. Americans are asking, who attacked our country? The evidence we have gathered all points to a collection of loosely affiliated terrorist organizations known as Al Qaida. They are some of the murderers indicted for bombing American Embassies in Tanzania and Kenya, and responsible for bombing the U.S.S. *Cole*. Al Qaida is to terror what the Mafia is to crime. But its goal is not making money. Its goal is remaking the world and imposing its radical beliefs on people everywhere . . .

Tonight, the United States of America makes the following demands on the Taliban: Deliver to United States authorities all the leaders of Al Qaida who hide in your land. Release all foreign nationals, including American citizens, you have unjustly imprisoned. Protect foreign journalists, diplomats, and aid workers in your country. Close immediately and permanently every terrorist training camp in Afghanistan, and hand over every terrorist and every person in their support structure to appropriate authorities. Give the United States full access to terrorist training camps, so we can make sure they are no longer operating. These demands are not open to negotiation or discussion. The Taliban must act and act immediately. They will hand over the terrorists, or they will share in their fate . . .

We are not deceived by their pretenses to piety. We have seen their kind before. They are the heirs of all the murderous ideologies of the 20th century. By sacrificing human life to serve their radical visions, by abandoning every value except the will to power, they follow in the path of fascism and Nazism and totalitarianism. And they will follow that path all the way, to where it ends, in history's unmarked grave of discarded lies . . .

Great harm has been done to us. We have suffered a great loss. And in our grief and anger, we have found our mission and our moment. Freedom and fear are at war. The advance of human freedom, the great achievement of our time and the great hope of every time, now depends on us. Our Nation—this generation—will lift a dark threat of violence from our people and our future. We will rally the world to this cause by our efforts, by our courage. We will not tire; we will not falter; and we will not fail.[31]

This second speech has a much more warlike character throughout its entire text. It made clear the president's position, that the United States would undertake any military action necessary to destroy al Qaeda, as well as any government standing in the way. Little did Bush know the enormous magnitude of the war that would follow and the fundamental changes in military technology, doctrine, and tactics that would emerge as a result of the fight at hand.

Predictably, the U.S. government, prodded by the public, had to pursue the capture or killing of bin Laden and his top lieutenants. They simply could not be allowed to launch an attack that killed thousands of American civilians without massive retaliation from the largest military power on the planet. It was no secret that the leaders of al Qaeda had accepted refuge from the Taliban government of Afghanistan. President Bush had clearly presented an ultimatum to the Taliban in his September 20 speech to Congress, when he stated:

And tonight, the United States of America makes the following demands on the Taliban: Deliver to United States authorities all the leaders of Al Qaida who hide in your land. Release all foreign nationals, including American citizens, you have unjustly imprisoned. Protect foreign journalists, diplomats, and aid workers in your country. Close immediately and permanently every terrorist training camp in Afghanistan, and hand over every terrorist and every person in their support structure to appropriate authorities. Give the United States full access to terrorist training camps, so we can make sure they are no longer operating. These demands are not open to negotiation or discussion. The Taliban must act and act immediately. They will hand over the terrorists, or they will share their fate.[32]At the same time, he asked the assistance of the rest of the world but also warned them of the dangers of supporting al Qaeda:

This is not, however, just America's fight, and what is at stake is not just America's freedom. This is the world's fight. This is civilization's fight. This is the fight of all who believe in progress and pluralism, tolerance and freedom.

We ask every nation to join us. We will ask, and we will need, the help of police forces, intelligence services, and banking systems around the world. The United States is grateful that many nations and many international organizations have already responded, with sympathy and with support, nations from Latin America, to Asia, to Africa, to Europe, to the Islamic world. Perhaps the NATO Charter reflects best the attitude of the world: An attack on one is an attack on all.

The civilized world is rallying to America's side. They understand that if this terror goes unpunished, their own cities, their own citizens may be next. Terror, unanswered, can not only bring down buildings, it can threaten the stability of legitimate governments. And you know what? We're not going to allow it.

Our response involves far more than instant retaliation and isolated strikes. Americans should not expect one battle but a lengthy campaign, unlike any other we have ever seen. It may include dramatic strikes, visible on TV, and covert operations, secret even in success. We will starve terrorists of funding, turn them one against another, drive them from place to place, until there is no refuge or rest. And we will pursue nations that provide aid or safe haven to terrorism. Every nation, in every region, now has a decision to make. *Either you are with us, or you are with the terrorists. From this day forward, any nation that continues to harbor or support terrorism will be regarded by the United States as a hostile regime.*[33] [emphasis added]

In this fashion, he sought to portray the coming conflict as one of good versus evil. This position played into the narrative being spread by bin Laden and other Islamic fundamentalists, who wished to convey the notion that the United States had set out not just to attack al Qaeda but to destroy

Islam itself. The common enemy had been created and was now clearly planning to invade a Muslim country on a mission of revenge, regardless of the costs or consequences.

In this speech, President Bush also began preparing the American public for the long-term costs associated with conducting a war on al Qaeda. Future terror strikes on American soil were entirely likely, and the conflict certainly would not be wrapped up in a matter of a few days. Unlike his predecessor, Bush was not content to fire a handful of cruise missiles or launch a series of airstrikes. Having suffered the loss of thousands of citizens, the United States was about to embark upon a crusade against the very concept of terrorism, and no nation would be permitted to remain neutral in the coming fight. Over the following decades, the newly declared War on Terror came to challenge the world's understanding of acceptable legal and moral practices in warfare, the technology best associated with modern conflicts, and even the concept of geographic boundaries traditionally placed upon warfare. Bush fully intended to pursue al Qaeda and its leaders to the ends of the earth, using any and all means at his disposal to destroy the terror organization.

RMAs have occurred on several occasions in human history, and they have tended to significantly upset the global power structure. As they permeate to new regions, they gradually become the new norm for human conflict—and areas or populations that refuse to adopt the new innovations tend to be victimized by those that accept it at the first opportunity. The primary contention of this work is that military robotics constitute just such a revolution and that those nations and populations actively pursuing them should be aware of the dangerous potential that they represent, rather than blindly driving forward in the quest for a deadly new technology. Of course, I am not the first author to contend that military robotics represent an RMA, nor am I the most alarmist when discussing their potential. Richard Falk argues "Weaponized drones are probably the most troublesome weapon added to the arsenal of war making since the atomic bomb, *and from the perspective of world order*, may turn out to be even more dangerous in its implications.[34] Grégoire Chamayou argues that military robotics are making wars more likely, in part because the wars of the future, at least for the nations equipped with such technology, might be fought with little or no losses in human personnel. According to Chamayou,

Using new means, the drone procures for its operators an even greater sense of invulnerability. Today as yesterday, the radical imbalance in exposure to death leads to a redefinition of relations of hostility and of the very sense of what is called "waging war." Warfare, by distancing itself totally from the model of hand-to-hand combat, becomes something quite different, a "state of violence" of a different kind. It degenerates into slaughter or hunting. One no longer fights the enemy, one eliminates him, as one shoots rabbits.[35]

Chamayou's point is well taken, in that the traditional conceptions of war put at least some risk upon both sides and that without such a risk, perhaps the term "war" might be misapplied to the violence being conducted.

However, not every scholar sees it in the same fashion—and some envision a conflict in which the human casualties of conflict might be reduced almost to zero. Barbara Ehrenreich, perhaps best known for her study of the origins of World War I, can hardly be called a warmonger—and yet, she sees the development of military robotics as a potential panacea, at least under certain circumstances, arguing:

An alternative approach is to eliminate or drastically reduce the military's dependence upon human beings of any kind. This would have been an almost unthinkable proposition a few decades ago, but technologies employed in Iraq and Afghanistan have steadily stripped away the human role in war. Drones, directed from sites up to 7,500 miles away in the western United States, are replacing manned aircraft. Video cameras, borne by drones, substitute for human scouts or information gathered by pilots. Robots disarm roadside bombs. When American forces invaded Iraq in 2003, no robots accompanied them; by 2008, there were 12,000 participating in the war. Only a handful of drones were used in the initial invasion; today, the U.S. military has an inventory of more than 7,000, ranging from the familiar Predator to tiny Ravens and Wasps used to transmit video images of events on the ground. Far stronger fighting machines are in the works, like swarms of lethal "cyborg insects" that could potentially replace human infantry.[36]

The truth, as is so often the case, falls somewhere between the extremes. Currently, military robots have been in wide-scale usage by the United States for less than two decades and by other nations for even less time. They represent an enormous potential, both for good and for evil, and the development of the technology itself is only one aspect of how they will influence the wars of the future. Military robots are likely here to stay, as they are simply too useful to be discarded, but there is still time to set some definite limits upon their employment.

CHAPTER 2

A Short Guide to Revolutions

[S]ince the discovery of gunpowder has changed the art of war, the whole system has, in consequence, been changed. Strength of body, the first quality among the heroes of antiquity, is at presence of no significance. Strategem vanquishes strength, and art overcomes courage. The understanding of the general has more influence on the fortunate or unfortunate consequences of the campaign than the prowess of the combatants. Prudence prepares and traces the route that valor must pursue; boldness directs the execution, and ability, not good fortune, wins the applause of the well informed.

—Frederick the Great, 1759

Military forces are inherently learning organizations, which are capable of rapid adaptation to changes in technology, doctrine, environmental conditions, and the behavior of enemies. They often constitute the first-adopters of new technological innovations, and in fact, most first-rate military organizations seek to produce technological changes that will supply an inherent advantage over potential competitors on the battlefield. In this regard, the field of military robotics, and the effect that it is having upon military forces around the globe, is no different from other revolutionary technological changes that have forever altered the nature of human conflict. Although the technology available at any given time in history has always been in flux, the changes in the modern era have come at an exponentially increasing rate, driven by both the raw number of individuals participating in research operations and the equally rapid rise in artificial computational power. Innovation in military robotics has a certain element that has not been present in previous military technological developments, in that the technology in question, robots, might actually serve to create more advanced versions of itself and might also be able to inherently improve itself through the incorporation of learning and modeling behaviors. To truly understand the unique nature of the rise of military robotics, though, it is necessary to examine the nature of advances in military technology throughout history.

REVOLUTIONS IN MILITARY AFFAIRS

In 1984, Soviet military theorist Nikolai Vasilyevich Ogarkov posited the notion of a revolution in military affairs (RMA), in particular, referring to the need for the Soviet Union to radically alter its approach to conflict if it hoped to remain in peer competition with the United States. While he might not have been the first person to suggest that certain technologies permanently altered the conduct of warfare, he at least deserves credit for coining the term that has become almost synonymous with radical technological shifts. Ogarkov pointed to a number of key examples of innovations that forced every major military on earth to either adapt to the new paradigm or fall victim to others who had embraced the new system of war.[1] Other writers built upon his broad concept and have proposed dozens of innovations as examples of an RMA. For some authors, the threshold for "transformation" has proven relatively low, while others reserve the term "RMA" to refer to only a few specific changes in human conflict. Regardless of the number of ideas that might be considered to qualify, an RMA is, briefly stated, a fundamental transformation in the means or methods of conducting warfare that conveys a massive advantage to adopters when engaged in conflict with non-adopters and which eventually establishes a new norm for the profession of arms. RMAs are not limited to technological changes, although many of the most commonly accepted changes had a major technological component. Examples include the adoption of gunpowder firearms, nuclear weaponry, or networked and computerized communications systems. However, the development of mass conscript armies, the industrialized production of munitions, or the use of combined forces of different types of specialized units (infantry, cavalry, and artillery, for instance) can also be considered RMAs with an equally important effect.

RMAs do not occur in a vacuum, nor are they instantaneous. The effects of a new idea might be immediately evident, particularly if it results in a decisive battlefield victory, but it also might take decades for the new concept to spread throughout the military profession. However, regardless of the speed of adoption, once an RMA has occurred, any military forces that stubbornly refuse to conform to the new concept or that prove incapable of such adaptation have tended to find themselves on the losing end of conflicts. Such advantages have often been the means by which empires rise and fall—a single key military innovation can have a cascading effect allowing the conquest of nearby rivals, increasing the power base of an aggressor state, and allowing further acquisitions of territory. To stop such a movement often requires peer competitors to adopt the same innovations, proving the adage "if you can't beat them, join them." Another way to state the same concept, one with much more frightening ramifications, is more commonly used in reference to the spread of religions by military conquest. It is simple: convert or die.

Fundamental changes in the nature of warfare are not a common occurrence, and, as the succeeding pages will demonstrate, it is possible for the art of war to appear stagnant for decades or even centuries. To the practitioners of war, that is usually not the case—they continue to innovate in every battle and campaign, sure in the knowledge that what worked today might never work again and that to become predictable is to court defeat. Not surprisingly, most changes in warfare are by definition evolutionary— gradual adjustments in strategy, doctrine, or technology in the hope of gaining an advantage and preserving it as long as possible. Thus, in the centuries after the fall of the Roman Empire, the knights of Europe gradually donned heavier armor, experimented with new weaponry, and tested new tactics in the almost-continual warfare that defined the era.

Trying to recount every RMA throughout human history is beyond the scope of this investigation, but there are a few that can serve as useful examples for illustrating how a fundamental change in the dominant form of warfare can have enormous effects. Not only do the adopters of a new RMA enjoy a significant advantage over those who have failed to adopt the new idea, they also have the opportunity to press that advantage for long-lasting political and demographic changes. States that embrace RMAs faster than their competitors create opportunities to overturn the status quo and potentially become a dominant regional or global power. The very desire to maintain the status quo should serve as a motivating factor for dominant states to seek out and embrace revolutionary ideas or technological changes—and yet, because those states maintain power under existing conditions, they also have a competing desire to avoid introducing any destabilizing factors that might threaten their own power base. Thus, it is essentially in the interest of dominant powers to remain aware of any potential RMAs and to either prevent their development by competitors or adopt the new concepts faster before they become an existential threat in the hands of the enemy. History is replete with examples of hegemonic powers that dismissed an RMA until it was too late to react to the new paradigm. Those hegemons that have maintained their dominance over long periods of time have typically been the same ones that have not categorically refused to consider changing their primary methods of applying military force in military conflicts.

"REVOLUTION"

The term "revolution" has been applied in many contexts, with both a positive and negative connotation. *Merriam-Webster's Dictionary* defines a revolution as "a sudden, radical, or complete change." In warfare, the modifier "revolutionary" has very specific implications. In general, revolutionary wars are a subset of civil wars. They may be waged by one portion of a nation attempting to break free of a central government, as was the case with both the American War of Revolution and the U.S. Civil War.

In those types of conflicts, the breakaway groups tend to view themselves as patriots freeing their territory from an abusive or oppressing foreigner, even if there is a long-shared heritage between the governors and the governed. Colonies casting off the imperial yoke often refer to the revolutionary aspects of their efforts to obtain independence, even if they manage to do so through political or diplomatic means. Revolutionary warfare can also be used to describe a conflict in which one segment of the population wishes to destroy or supplant the existing government with a radically different form of governance. Examples of this style of warfare include the French and Russian Revolutions, both of which removed by force a monarchy in favor of a more republican style of government. The French effort lumbered through a number of revolutionary governments, each replacing its predecessor in a bloody wave of violence, before being effectively terminated through the coronation of Emperor Napoleon Bonaparte a mere 15 years after the revolution began. He, in turn, held sole power for less than 10 years before a coalition of external military forces, each under a monarchical government, removed him and restored the Bourbon family to power. The Russian Revolution removed the tsars, to be sure, but proved no more successful than the French experiment, as the short-lived Russian government run by Alexander Karensky quickly lost any semblance of control. A bloody civil war between the Bolsheviks and the Mensheviks solidified the establishment of a new communist government under Vladimir Lenin. For any American readers feeling smug about the U.S. experience, it would be wise to examine the number of internal revolts against the new republic in the aftermath of the 1783 Treaty of Paris.[2] Likewise, it must be noted that the revolutionary U.S. government fared no better than its European counterparts and proved completely incapable of defending its own territory. Not until after the ratification of the U.S. Constitution and the election of George Washington as a reasonably powerful executive did the federal government show even a semblance of successful function.

A revolution can also be used to signify fundamental changes in other elements of human society. For example, the Industrial Revolution, which commenced in Britain in the late 18th century, completely altered the methods of production around the globe—and any nation that refused to accept the new systems of manufacturing could expect to be economically bypassed by all competitors who did. Over time, most nations adopted the tenets of the industrial process or at least did not attempt to compete in manufacturing by more traditional means. Nations with better access to the key resources for industrialization naturally had certain advantages, as did those geographically well situated to exploit access to large consumer markets. Yet, even some nations with few industrial resources, such as Japan, have built themselves into industrial powerhouses through the sheer determination to do so. Not coincidentally, many of the same early adopters of industrialization have proven most open to utilizing the advantages of robotic developments in their manufacturing centers, demonstrating a continuing willingness to advance their industrial economies and maintain their economic primacy in the world.

Many have dubbed the creation of the Internet and the massive increase in the number of communications avenues an "information revolution." This fundamental shift in the means of transmitting data has had far-reaching, and often unexpected, social and political effects. It has made controlling information far more difficult for repressive governments, as was demonstrated by the use of social media to coordinate 2012's Arab Spring. It has also created an amazing opportunity for intelligence agencies, which have wholeheartedly embraced using computer networks as a means to penetrate classified, sensitive programs. Not only do cyber networks provide the skilled attacker access to key data, they also enable its exfiltration, dissemination, and re-encryption.

Most RMAs occur over a period of years or even decades, which can make pinpointing the exact moment of change almost impossible. Yet, for all of their apparent gradual change, there have always been winners and losers as a result of every RMA, and differentiating between the two is usually quite simple. Often, the winners are those who survive the transformation and use it to further their own interests and position within the international arena. The losers are fortunate if they manage to avoid annihilation; preserving their previous standing has always proven impossible except when they had recognized the RMA and enthusiastically adapted to the changed nature of conflict. Regaining past glory is exceedingly rare in international relations—once an empire has collapsed, the remnants of it may remain, but they are unlikely to achieve a status on par with the original imperial position.

Change is difficult for states and embracing a fundamental transformation even more so. When it comes to RMAs, it has proven far easier for weaker states to accept the need to change their approach to warfare, if only to have a better chance at survival. Tim Harford makes the case that success always comes after failure, because only by learning from mistakes can an organization truly open itself to the need for change and undertake the necessary actions to support it. In particular, he sees adaptation to changing circumstances as the most important attribute of a leader.[3] Although he focused more upon businesses than nation-states, and his concept should not be considered an absolute requirement, in the case of RMAs, early adopters have often sought to obtain an advantage over stronger rivals by testing a new technology, doctrine, or strategy. Harford summarizes his primary concept in what he calls the "Palchinsky Principles," namely, to try new things, to do so on a scale small enough to survive potential failure, and to learn from mistakes.[4]

Gaining a short-term advantage means little if one cannot translate it into a lasting position of strength. Weak states are often weak due to factors beyond their immediate control, such as a lack of defensible borders, a dearth of natural resources, or an inability to participate in useful trade of goods and ideas with neighbors. While such a state might manage to embrace change in warfare, it will have a very limited period to take advantage of any innovation before its more powerful neighbors adapt to the new concept, either by countering it or by adopting it for themselves. As an RMA permeates a region and expands, its competitive advantage dissipates and is useful only against an opponent who has not embraced the new concept. At that point, a new equilibrium develops, which remains the norm until another disruptive innovation comes along to once again threaten the balance of power.

To a certain extent, the more disparate the technological development levels between two combatants, the more obvious the improvements from an innovation will appear. When matchlock muskets first appeared on the battlefields of Europe, they were a significant improvement over earlier musket designs but not so much that they conveyed an overwhelming advantage over those who continued to use the earlier designs. When those same matchlock musketeers first took the field against African, Asian, or American warriors armed with hand weapons, they provided a devastating advantage that could not be easily overcome even by an enormous numerical advantage. European colonial powers swept around the globe in large part due to their superior armaments, and even relatively weak European states like Portugal and the Netherlands were able to construct enormous empires because they possessed much greater military capabilities. Once again, though, being first to the game proved insufficient to guarantee a lasting advantage over more powerful rivals. Despite moving much later into colonialism, the British and French parlayed their advantages in Europe into the two largest colonial empires, gradually eclipsing their earlier competitors.

A prominent and undeniable example of an RMA in action is the easiest way to illustrate the concept. When gunpowder weapons became prevalent in Europe, the new technology offered a significant advantage for early adopters, who could use its explosive power to batter down fortifications from afar. However, even gunpowder could not offset the population differential between Switzerland and France—even if the Alpine kingdom had adopted firearms first, it had little hope of conquering its much larger neighbor, and any attempt to do so would require sending a small expeditionary army beyond the defensible Swiss borders on a mission to attack the most powerful state in Europe. There is simply no conceivable way for such an action to result in anything but defeat. Had firearms allowed the Swiss to destroy a French army, French commanders would likely seek to avoid battle while also working to capture or counter the new technology. French armies with pre-gunpowder weapons would still represent a threat to the Swiss, if only through their massive numerical superiority.

In the 21st century, the speed of RMAs seems to be increasing, perhaps in part due to the exponential growth of the speed and capacity of information systems. Although computers might not constitute an RMA in and of themselves, their existence undeniably enables other RMAs to occur. In large part, this is due to the compression effect created by massive information storage and transferal capabilities. No longer would it be necessary to gather the finest minds in physics at a single location in the New Mexico desert to pursue a Manhattan Project, although there still might be secondary benefits to that approach. Scientific collaboration is now possible from anywhere in the world, and the development of encryption systems mitigates most, though not all, of the security-related fears that permeate the national defense establishment. Likewise, because the entire world can see the effects of technological developments as they are being fielded, there is certainly a better option to copy systems and behaviors than simply waiting until they are applied against your own nation.

There is a long history of human societies establishing boundaries upon military innovations and operational employment of new technology. Some of the historical limits were established through formal legal mechanisms, while others were devised through common cultural understandings and informal agreements or on the basis of sustained moral arguments. Still others were essentially created through a mutual fear of retaliation, particularly if both sides possessed, or could quickly obtain, a dangerous new form of weaponry. Some limits have pertained to specific technological advances, such as the attempt to limit the proliferation of nuclear weapons, while others have referred to behaviors rather than the tools used. In this category, prohibitions upon targeting civilians or executing prisoners of war serve as applicable examples.

THE EARLIEST RMAS

There is a case to be made that the earliest and greatest change in the nature of human fighting occurred long before the development of a written language to record the event and probably even before anything approximating human speech existed to describe the event. The existence of conflict among humans is undeniable, though, and has long been perceived as a uniquely human attribute.[5] That happened whenever a primitive human first picked up a rock, or a stick, or some other primitive tool readily at hand and used it to attack another human. The advantages of such a simple weapon were too great to ignore, and long before humans organized into societies capable of conducting anything that might be considered warfare, they had learned to custom-design weaponry using the materials readily at hand. Fire-hardened spears were soon augmented by flint rocks knapped into razor-sharp points. Rawhide, sticks, and smooth rocks made for extremely accurate slings. Atlatls greatly augmented the range of early missile weapons, as did simple bows and arrows. In short, one of the earliest forms of human technological development came in the creation of rudimentary weapons, which served as tools for hunting and also as a means of fighting other humans.

Ancient eras of human civilization are often described in terms of technological prowess, and in particular, the key resource used in tool construction, including weapons. Thus, stone-age cultures gave way to those using copper, which could be formed into a form of armor and much sharper weapons. Bronze toolmaking supplanted the use of copper, as bronze tools held their edges much better and created far more durable weapons and armor. In turn, societies that developed ironworking could overwhelm bronze-using civilizations in short order, and those that converted to steel had a marked advantage over those still languishing in the Iron Age. These advances did not happen in an instant; the knowledge of how to work different materials required generations of experimentation before it permeated a given society. However, once a new technique was proven superior to the existing state of the art, it became a closely guarded state secret. A major technological advantage also created an opportunity for outward expansion and domination of nearby rivals, one which lasted, however, only until all of the powers in the region had adopted the new innovation.[6]

THE PHALANX: A REVOLUTIONARY FORMATION

To truly see the effect of changing military innovations, one needs only examine how ancient empires formed, often on the basis of a single revolutionary idea. Thus, for example, Greek city-states of the classical period, whose armies relied upon relatively simple long spears, handheld shields, chariots, and body armor, managed to spread throughout the eastern Mediterranean basin and as far west as Sicily and Carthage. The fact that Greek hoplites carried such weapons did not revolutionize warfare, although their weaponry and other tools of war were as advanced as anything in the immediate vicinity. Rather, their strength came from the methods they used when deploying their technology, including the tactical decision to operate in a formation called the phalanx and the social decision to rely upon the concept of citizen-soldiers. The phalanx allowed the Greeks to move across the battlefield in a formation that maximized the protection of each member, in part by requiring soldiers to protect one another with their overlapping shields. From behind this wall of protection, Greek troops could thrust their spears against the enemy. The phalanx was not particularly nimble, but it did present a seemingly unstoppable force that destroyed any units foolish enough to stand against it. Even against much larger enemy forces, the phalanx presented a significant tactical obstacle, so long as its flanks could be kept secure. The vaunted 300 Spartans used a phalanx formation (and approximately 8,000 auxiliaries from other Greek city-states) to hold off an enormous Persian force at Thermopylae in 480 BCE.

The phalanx had a signal weakness, as is often the case with seemingly invincible formations. It moved slowly and almost always in a straight line. If it could be approached from the flanks, it could be broken. Likewise, troops operating in the phalanx had little chance of chasing down an enemy that sought to avoid battle; thus, it could be employed only against an enemy willing to meet on a battlefield. Soldiers in the phalanx had little chance of successfully attacking a fortified position, such as a walled city, although they might successfully carry out a siege by surrounding and starving the enemy out. They could possibly take the city by storm, but doing so tended to be costly in blood and treasure and certainly did not occur within the tactical formation in question.

The phalanx proved so effective that it became the dominant form of infantry combat formation for centuries, with only minor modifications. It had a significant advantage in that the function of each member within the unit was relatively simple, meaning it did not require the fulltime training and dedication that would be afforded to professional troops. This allowed Greek city-states to mobilize a large percentage of their adult males when required by crisis, without the added expense of supporting a standing army. Only one major city-state, Sparta, deviated from the Greek norm of the citizen-soldier. Unlike its rivals, Sparta had conquered and absorbed a nearby city-state, Messenia, and used the production of its conquered vassals to offset the costs of a professional fighting force (which was largely necessitated due to the threat of a revolt from those same vassals). Spartans devoted an inordinate amount of time and effort to perfecting their martial skills and in time became the warrior elite of the region. Yet, even Spartans, for all their prowess, could not devise a means to overcome the phalanx beyond forming their own, similar formations and facing the enemy in close combat. Further, Spartans hesitated to campaign far afield from their homes, as they continually feared a slave revolt if they marched too many troops away for a long campaign.

The Peloponnesian War (431–404 BCE) demonstrates all of the advantages (and the few disadvantages) of the phalanx system. When Athens and Sparta came to blows over hegemony in the region, it pitted a naval power (Athens) against a land power (Sparta). Athenian troops could not defeat their rivals on the battlefield—but soon realized they did not need to do so in order to win the war. Instead, the Athenians built enormous walls to protect their city and its port, Piraeus, as well as the connecting road between the two. The Athenian navy kept the city supplied and ensured that any Spartan siege efforts would not only be in vain but would also expose their own city to a seaborne assault. Not until the Athenians engaged in a ruinously expensive attempt to conquer Syracuse did the Spartans have much opportunity to inflict a decisive defeat, but even that humiliating failure did not induce Athens to surrender. Eventually, the Spartans allied themselves with ancient foe, Persia, which supplied a large enough navy to offset the Athenian advantages at sea, gradually leading to the fall of Athens. For all its power in the open field, the phalanx simply had no capacity to breach a strong fortification, and its slow movements ensured that any enemy that chose to flee would have little difficulty escaping. Thus, an outmatched enemy could choose to fight or flee, and the phalanx would essentially have to accept that decision.[7]

The phalanx was a perfect representation of the democratic ideal held by many of the Greek city-states. Within its ranks, each member was essentially equal, and the entire body of the formation could function only if its component parts acted in concert. Any member who failed to perform his assigned role created a gap within the formation, one that could be exploited by an enemy. In much the same fashion, the political power of an individual city-state required its members to present a unified position once the internal debates had concluded. There was simply no tolerance for a divided polity in time of war—either the entire society mobilized to support a conflict or it could expect to fall to the enemy with all of the dire consequences that were likely to follow.[8]

Another interesting facet of the phalanx was how easily new members could be added to the formation. There was no requirement for years of training, or even for complex tactical decision making for the vast majority of its members. Participants within the phalanx had a very well-defined role that depended entirely upon their position within the formation. Those in the front ranks had the most important defensive function, as their large shields served to protect the entire unit from direct-fire projectiles. Each member's shield coverage overlapped the body of the man to his left, and each member was protected in turn by the man to his right. Only the ends of the line had a different situation; otherwise, everyone in the line served in essentially the same fashion. Those behind the front rank thrust their spears forward over the men in front of them, seeking vulnerable points in the defenses of any enemy in range. The back ranks also bolstered those before them by placing their shields on the backs of the rank immediately before them and pushing. In effect, the phalanx became a grinding, almost unstoppable wall of flesh and metal, pushing back or rolling over the enemy. Those who fell before it were simply crushed under its weight, pressed down into the earth before being stabbed by one of the follow-on ranks.[9]

So, what were the battlefield vulnerabilities of the phalanx? For a time, it seemed almost invincible, particularly when matched against an enemy employing a different formation. On numerous occasions, Greek phalanxes triumphed over much larger but more chaotic formations, as at the Battle of Marathon in 490 BCE, when approximately 10,000 Greek hoplites decisively defeated a Persian army of at least 25,000. Yet, the phalanx had an inherent weaknesses. Its slow speed and dense formation made it an ideal target for even the most rudimentary of siege weapons. While a hoplite's shield might serve to stop ordinary arrows fired by archers, it could do little to stop larger missiles hurled by even crude engines. The phalanx also struggled when faced by heavy shock cavalry, such as that employed by Philip of Macedon. Philip's son, Alexander the Great, proved even more adept at using heavy cavalry to break up enemy formations, which then left them open to attack from his own infantry.[10]

The Roman Empire was built in large part through the creation of a massive, extremely capable military force. Its infantry units resembled phalanxes on the surface, in that they operated in linear formations that relied upon unit cohesion as a key determinant in a battle's outcome. Each Roman soldier carried a long semicylindrical shield, the scutum, that offered protection almost from head to toe. However, rather than using a long, heavy spear for attacking, Roman troops relied primarily upon pilae and short swords. Pilae were javelins with long iron blades. They could be flung up to 50 meters, putting enemy forces in significant danger before they could come close with Roman units. Each pilum was designed to bend at its neck upon striking a solid object, whether it was an enemy shield or the ground. Thus, it could not simply be picked up and flung back at the Roman forces, and if it lodged in a shield, it became exceedingly difficult to remove, particularly under battle conditions.[11] An enemy without a shield fell quickly to Roman archers interspersed within the heavy infantry units, immediately tearing a hole in the formation. When Roman generals gave the order to advance, their legions advanced just as inexorably as the earlier Greek phalanxes, but the reliance upon the short thrusting sword offered a quicker and more nimble form of killing in the front ranks. Trailing lines still pushed upon the front line, driving the entire formation into the enemy and seeking a breakthrough, which typically caused the enemy to break and flee, leaving themselves open to being cut down from behind by the Roman cavalry waiting on the flanks.

Roman military dominance did not rest solely upon creating an improved infantry formation. It also relied heavily upon engineering knowledge, a hallmark of Roman civilization. Roman forces could be counted upon to build fieldworks in a very short period, offering protection to campsites and making it very difficult for an enemy to take Roman legions by surprise. Their engineering capabilities also made Roman armies able to undertake sieges of strongly fortified positions, something their Greek forebears struggled to accomplish. Engineering prowess also enabled the construction of extensive logistical systems to supply ever-increasing field armies flung along the frontiers of a constantly growing empire, largely by building roads that have remained usable for two millennia.

The Roman Empire eventually collapsed, a downfall hastened by continual attacks upon its periphery. When it fell, its approach to military operations also disappeared, to be replaced by a more mobile style of warfare consisting of mounted bands of heavily armed warriors. Over time, European states became increasingly reliant upon cavalry units backed by lightly armed infantry. It took centuries for the mounted knight to become the symbol of European warfare, as well as the noble class commanding the commoners. Heavily armored knights accepted responsibility for the military activities of the state and reaped the resulting rewards by assuming the rule over the citizenry. This professional class of the warrior markedly differed from the earlier Greek and Roman soldiers, but like them, well represented the society that produced him. The European feudal society included a complex system of pledging fealty to the more powerful in exchange for the protection they could provide. In return for a subject's loyalty, support in wartime, and provision of a predetermined number of troops, a liege guaranteed the safety and sovereignty of his vassals. Like previous civilizations, this system functioned largely in accordance with the harvest cycles, as the bulk of European troops were not fulltime warriors so much as seasonal combatants.

One early example of this phenomenon was the crossbow, a relatively simple-to-operate device that could be introduced to new recruits in a matter of a few hours of training. As handheld crossbows became more powerful, they possessed enough penetrating power to drive a bolt through heavy plate mail armor. Thus, a poorly equipped peasant newly conscripted into the military might represent a mortal threat to a heavily armored knight who had trained for war his entire life. The knights, who served as the symbol of European nobility, could not tolerate such a calamity and the social upheaval it might create. To counter this new and terrible weapon, they appealed to Pope Innocent II, who formally banned the use of crossbows in "Christian warfare" as part of the Second Lateran Council in 1139. Of course, this prohibition did not apply to wars against heathens, and the ignorant peasants were free to fire entire quivers of bolts at any armored non-Christians they might wish to target. However, at least in theory, the nobility of Europe could stay safe from all combatants but fellow elites, so long as they professed an adherence to Christianity.

The crossbow example is poignant, because these ranged weapons continued to increase in sophistication and power in spite of the Pope's attempts to limit them. By the late 12th century, the windlass crank allowed a crossbow-armed soldier to load an 800-pound draw weight crossbow. Naturally, the English archers could maintain a far higher rate of fire than the crossbowmen equipped by rival nations, a fact amply demonstrated on the battlefields of Crecy, Poitiers, and Agincourt, but they represented little threat to the heavy cavalry favored by the Continental powers. Not surprisingly, many European military leaders found ways to justify their use of crossbows against fellow European Christians—the device was just too effective to be ignored.

When knights faced a non-European military force, their strengths and weaknesses appeared in stark contrast. Thus, for example, when European warriors invaded the Levant, motivated by religious fervor or the expectation of riches, they met enemy forces with a radically different conception of how to wage war. The lightly armed Arabs had little chance of standing up to a heavy cavalry charge—and they quickly learned not to try. Instead, they fought a war of maneuver, darting close to pepper the Europeans with arrows and ducking away before a melee ensued. The Europeans could march to any fixed point and sack it, a lesson amply demonstrated by the siege of Jerusalem in 1099 during the First Crusade. Holding these fixed positions, particularly those not on the Mediterranean coast, proved a far more difficult proposition. The Europeans' logistical system proved inadequate to the task, and supply trains fell victim to raiders long before they reached their destination. By 1187, Muslim commander Saladin was able to force upon the remaining Crusaders in Jerusalem the choice of surrender or death. In 1291, the Crusader presence in the Levant ended with the fall of their last outpost at Acre.

The Crusades proved a series of costly undertakings for European monarchs, but they did have the undeniable benefit of pulling some of the most bloodthirsty elements out of society and sending them over the sea, many of them never to return. By the end of the Crusades, some significant military changes had started to appear in Europe. One was the return of missile weapons as a key component of warfare. Archery had never disappeared completely, but bands of archers presented little danger to heavily armored knights. However, by the 12th century, a new form of archer became increasingly common in Europe: the heavy crossbowman. Armed with a powerful spring-assisted bow, these warriors, as already noted, could fire a projectile with sufficient force to pierce all but the heaviest plate armor. They had the added benefit of being relatively cheap to equip and easy to train, as the mechanics of crossbows required only a few minutes to learn. When firing by volleys, even aiming was a relatively optional activity. Although Pope Innocent II expressly banned its use against Christian enemies during the Lateran Council of 1139, his edict fell on deaf ears, as the weapon was simply too effective to be abandoned, even in the name of preserving the social order of the feudal knights.

Virtually every Continental military force incorporated units armed with crossbows, and mercenary companies specializing in the weapon quickly became common. As interest in crossbows grew, their designs became much more complex, using mechanical advantages to increase the power of individual weapons. Through innovations such as the windlass, crossbows could be produced with draw weights of several hundred pounds, far greater than the draws of the heaviest longbows ever discovered.[12] Knights had a finite limit to how much armor they could pile on, even if they remained on horseback, and those who wore the heaviest plate mail became exceedingly vulnerable if they could be unhorsed. Only the British Isles clung to the conventional hand-drawn bow. Not only did they continue to rely upon the older technology, English Common Law required that every adult male possess and be proficient in the use of a longbow.

To the modern layperson, there probably appears to be little difference between the longbow and the crossbow, as both seem hopelessly outdated in an era of automatic gunpowder weapons. In the 14th and 15th centuries, though, these weapons were state of the art and each possessed significant advantages. The choice of which to utilize might make the difference between victory and defeat, making it a literal life-and-death decision. Conventional bows had a far greater range, in part because arrows were designed to be more aerodynamic than bolts. Arrows had a wider variety of points, from the narrow bodkin point designed to penetrate chain mail armor to the razor-sharp flathead, created to slice through unprotected flesh and leather armor. A well-trained archer could fire at least 10 arrows per minute and had a realistic chance of striking a man-sized target at 100 meters or more. Crossbows had greater raw power, but to achieve it, they sacrificed speed, with the heaviest crossbows requiring two minutes between shots as the bowspring was painfully cranked into position. Although the crossbows were more difficult and expensive to produce than longbows, training troops to use them was a far easier process. Also, the truly paranoid ruler might take comfort in the knowledge that angry peons could not readily produce new crossbows on their own, while new longbows could be built in relatively short order.[13]

During the Hundred Years' War (1337–1453), the two technologies came into direct conflict. In the three most decisive battles of the conflict, English longbowmen outdueled their Continental crossbow-wielding opponents. This led to English victories at Crecy, Poitiers, and Agincourt, despite the English being outnumbered in each engagement. However, despite these victories, the English could not overwhelm the French and were eventually driven off the continent by the end of the war. Over the span of the 116-year conflict, it should be unsurprising that both sides were willing to test any new ideas that might bring about an advantage, including the utilization of a dangerous new chemical produced in extremely limited quantities: gunpowder.[14]

GUNPOWDER: THE MOST ACCEPTED RMA

Although gunpowder was relatively well known in China as early as 1000 CE, it remained somewhat of a novelty until it was brought along trade routes to Europe in the 13th century. In Europe, gunpowder was soon utilized as a powerful explosive used to propel rudimentary missiles with extremely high kinetic force. The earliest firearms were extremely difficult to load, notoriously unreliable, and painfully inaccurate. Although they were incredibly effective at certain aspects of warfare, this effectiveness was not so inevitable as to provoke an immediate change. The adoption of gunpowder was a process that required centuries to displace a style of warfare that had itself developed over more than a thousand years. By the time gunpowder had fully replaced its predecessors, though, it had not only changed the conduct of violence on the battlefield, it had destroyed the entire social order of Europe, sounding the death-knell of feudalism and triggering the rise of the modern nation-state.

The earliest gunpowder weapons fell into two categories: artillery and personal firearms. The artillery pieces consisted of hollowed logs or crudely cast iron and bronze tubes, into which a quantity of relatively primitive gunpowder was loaded. A projectile, usually either a hand-shaped stone or an iron javelin, was loaded directly on top of the powder, and an oil-soaked piece of rope was then touched to a firing-hole at the base of the weapon to detonate the charge. These missiles required little imagination or adaptation—both stones and javelins had been fired from mechanical siege weapons such as trebuchets and ballistae for more than a millennium. Use of these artillery pieces was an extremely dangerous activity, as they had a deadly habit of bursting, maiming, or killing their operators. However, when they functioned correctly, the results could be awe inspiring. A stone ball fired from a gunpowder weapon could shatter the high curtain walls of a traditional castle and could do so from a much greater range than any of its mechanical competitors, tearing great ragged holes in fixed fortifications that could then be breached by attacking infantry forces. The castles of old, which represented the highest art of military construction in their day, quickly became obsolete artifacts that could no longer perform their primary function of defending the forces within them, much less the surrounding population.

Personal firearms, which had much less utility in siege warfare, proved equally devastating on the battlefields of Europe. The earliest "hand cannons" consisted of short bronze tubes attached to wooden handles, essentially creating field artillery pieces in miniature form. They had to be loaded before the bearer entered the battle and essentially could not be reloaded in the midst of the chaos of a melee fight, making them a single-shot, fire-and-forget weapon. The effective range of these small firearms was probably less than five meters, and they might have been more effective at panicking enemy cavalry horses more than inflicting significant casualties upon the enemy. However, when fired, their missiles could easily punch through even the heaviest of plate armor, allowing the lowliest conscripted peasant to present a deadly threat to the mounted knight. Suddenly, the arms and armor possessed by the elite fighters of Europe and well beyond the reach of most combatants did not offer sure protection against the masses, and in some cases, became an expensive and ungainly encumbrance.

The gunpowder revolution conveyed a massive advantage to armies that adapted to the new conditions of warfare and rendered obsolete many of the previous forms of conflict. States in Europe that could not, or would not, adopt the new weaponry proved incapable of competing with the states that did accept the changes. Further, as European nations began to engage in voyages of exploration to other continents, they used their gunpowder weaponry to conquer the civilizations that they encountered. Even when European traders sold firearms to natives of other lands, they typically closely guarded the method of producing gunpowder, limiting the utility of firearms to any society determined to resist European advances.

The adoption of gunpowder is one of the most universally recognized RMAs in history, and an examination of the development of armies based upon gunpowder illustrates many of the characteristics of most RMAs. Adoption of the new capability extended over several decades of warfare, and its initial utilization did not immediately alter the organization or behavior of armed forces. Over time, though, it became increasingly central to almost every military operation and consideration, such that by the time gunpowder armies had matured, they could dominate any previous type of military organization that had not adopted the use of gunpowder weapons. When European colonial forces began their expansion into the Americas, sub-Saharan Africa, south Asia, and the Pacific, their possession of gunpowder technology offered them an insurmountable advantage against opponents without access to the same weapons. In many ways, gunpowder enabled European societies to dominate the globe and offered the opportunity for even relatively weak European states to develop enormous colonial empires.

Gunpowder was first created in China during the Song dynasty, probably approximately 1000 CE. The earliest forms of gunpowder relied upon the careful mixing of sulfur, saltpeter, realgar, and honey. Although combustible, the earliest versions of gunpowder did not offer the explosive force of gunpowder manufactured by more sophisticated techniques, and its creators did not immediately grasp the potential military utility of their invention.[15] Over time, experiments with different manufacturing processes and formulas refined Chinese gunpowder. When placed in a tube with one end capped, Chinese gunpowder supplied an effective propellant, allowing for the creation of rudimentary signaling rockets. When mixed with other powdered minerals and detonated, gunpowder could create a variety of colorful explosions, leading to the development of fireworks displays. On at least a few occasions, gunpowder bombs were used to breach fortifications or were flung via siege engines into enemy defenses. Yet, the development of an effective Chinese infantry weapon using gunpowder did not occur, perhaps in part because the relative stability of the Song dynasty, and its successors, did not necessitate major efforts in military innovation. The Mongols, on the other hand, embraced the explosive power of gunpowder, which, after their conquest of the Song dynasty in 1279, probably became a key vector for its expanded use. Gunpowder might have remained simply an Asian specialty for an indefinite period of time had it not come to the attention of European traders who established contact with the Chinese at the end of the 13th century.

It would not be fair to say that exposure to gunpowder immediately inspired the Europeans with visions of martial conquest, although it certainly surprised and delighted at least some of the traders, who brought samples back to their homes. Had the Chinese successfully attempted to prevent Europeans from learning the process of making more of the substance, history might have unfolded in a radically different fashion. However, the traders did not merely obtain small quantities of gunpowder—they also soon learned how to take the raw materials and combine them in the correct ratios, ensuring that the full transfer of gunpowder technology would soon occur.[16]

Just as gunpowder could propel a rocket forward, it could also be used to send a smaller projectile flying out of a capped tube, at an extremely high velocity to boot. Early European experiments with gunpowder weapons largely revolved around the development of rudimentary artillery. In part, this reflected the state of manufacturing in Europe, which had little capacity for the sophisticated craftsmanship required to produce handheld firearms. Working on a larger scale was simply an easier proposition. It also demonstrated the current state of European warfare—mounted, heavily armed and armored knights served as the primary symbol of European warfare. Heavy cavalry charges, like the phalanx of centuries earlier, could essentially sweep aside any other formation that might be thrown against them, ensuring that they would be unstoppable by anything short of a similarly equipped, trained, and organized force. Also like the phalanx, massive fixed fortifications proved extremely effective at defending against field armies—and nearly two millennia of engineering practice had created enormous, almost impregnable fortifications. Medieval sieges, even if successful, were measured in months, or even years, and the majority of sieges failed to capture their objective. Even towering siege engines flinging enormous boulders had little chance of battering down an enemy's walls. Their rate of fire and accuracy left a great deal to be desired, and they were susceptible to sallies from the defenders. In order to be within range of the enemy walls, they had to be constructed within a few hundred yards of their target, which not only made their erection obvious, it also brought them into range of similar engines manned by the defenders, which typically had greater range due to their elevated positions along the walls. Even without counterfires, the close range meant any unguarded siege engine was likely to be destroyed by a quick strike from the defenders, who would rush back to the protection of their walls before they could be caught in the open.

Gunpowder artillery could be constructed out of range and even sight of its target. It required far less physical space than a mechanical siege engine, which meant more gunpowder weapons could be brought to bear upon a single point of the enemy defenses. Even its earliest, most rudimentary designs demonstrated a far greater accuracy than that of mechanical systems, and its projectiles, which could be fired on almost a flat trajectory, struck with far greater kinetic force than that of their more primitive competitors. Although many early artillery pieces showed extreme vulnerability to accidents, often in the form of an explosion in the barrel that destroyed the piece and killed or wounded anyone standing nearby, the raw destructive power of gunpowder artillery could not be denied.

There was little agreement upon the best way to construct or employ gunpowder artillery upon its first introduction to European battlefields. Some early adopters experimented with hollowing logs to create wooden cannons. These field pieces were cheap and easy to construct but definitely could not be relied upon for more than a few shots before they fractured or caught fire. Efforts to band the logs with straps made from bronze or iron might extend the life of the field piece somewhat but also increase its cost, its construction time, and its weight. Some early adopters considered using layers of tanned leather, dried and hardened into tubes, on the assumption that it was less likely to ignite or burst. This method of construction offered much lighter field pieces but proved costly and susceptible to warping under the extreme stresses of containing a gunpowder explosion. It seemed that only a metal tube will offer the strength necessary to create an artillery piece that could survive the sustained rates of fire necessary to demolish the high curtain walls defending most European strongholds.[17]

The construction of metal artillery pieces made the utilization of gunpowder artillery almost the exclusive province of powerful political entities, at least in its earliest period of adoption. A great amount of trial and error commenced throughout Europe, with new designs continually appearing and being tested in the almost-constant warfare between European states. In short order, the time required to besiege a traditional fortress plummeted—modern historians can quickly demonstrate how sieges dropped from endeavors requiring multiple campaign seasons to relatively commonplace affairs lasting for only a few days. Thanks to gunpowder artillery, standing behind fixed defenses not only ceased to be a safe plan, such an idea became almost suicidal, as the enemy would require very little thought to determine one's position, surround it, and batter it into submission.[18]

Gunpowder offered the opportunity to completely upset the European balance of power. Prior to its introduction, one of the dominant powers of the region was the Byzantine Empire, whose capital, Constantinople, had survived for more than a millennia thanks in large part to its static defenses. When Attila the Hun ravaged Europe, including the remnants of the Roman Empire, the wealthiest city of the era, Constantinople, withstood his advances. When nomadic Arab tribes, inspired by Mohammed and the foundation of Islam, swept through the Middle East, North Africa, and parts of Europe, Constantinople proudly stood, unconquered and invincible. It survived dozens of threats from would-be conquerors, including a handful of protracted sieges from truly determined enemies. However, when Ottoman Sultan Mehmed II arrived at the gates of Constantinople in 1453, he brought an entirely new capability along with his army: bronze artillery. The proud city, which had withstood centuries of assaults, fell to the bombardment of his cannons in just eight weeks, boldly signifying the fundamental shift that had occurred in the dominant mode of warfare.[19]

Early handheld gunpowder weapons looked remarkably similar to a miniaturized version of artillery. They involved a metal tube that held the powder and the projectile and had a small touch-hold drilled at the base through which the powder charge could be ignited. They were often mounted on a short wooden handle, both for ease of use and for the safety of anyone attempting to employ them. Users had to touch a lit match (a length of rope dipped in oil and lit on fire to produce a smoldering flame) to the hole in the side of the firearm, which then ignited the main charge and fired the weapon. The effective range of such weapons was probably measured in feet, making them suitable for a surprise attack in a melee situation, but completely worthless for any degree of ranged fire. However, when used up close, the effect proved devastating—the sound accompanying the explosion might induce panic in any horses not trained to withstand its shock. The projectile fired from the gun could punch through even the heaviest armor and shields, and, because early bullets were round rather than pointed like crossbow bolts, they tore gaping wounds into human flesh. They also tended to carry pieces of the victims' clothing into the wound, making infections a deadly secondary effect from any gunshot wound. The projectiles also tended to lodge within the victim and, unlike arrows, were far harder to locate and remove without creating a substantial amount of additional injury. In short, even combatants who survived the initial shock, trauma, and blood loss from a gunshot wound still faced a much higher risk of dying from their combat injuries than those wounded by more traditional weapons.

Handheld firearms soon improved, with more reliable firing mechanisms, longer barrels for more accuracy, and standardized metal projectiles. Due to its low melting point and relative availability in Europe, lead soon became the most common material for bullets, which could be quickly cast by armies in the field using simple molds. Gunpowder itself also improved, with innovators testing different formulas and production systems to create more uniform and dependable supplies. Although gunpowder still remained susceptible to moisture, firearms were the primary armament of European armies by the early 16th century, just as the first major wave of colonization efforts began to take shape. Of course, even though firearms offered a means to penetrate heavy plate armor, they also had a number of drawbacks, particularly in the early models. The weapons were difficult to manufacture, as was the powder that fueled them.[20] Reloading early gunpowder weapons required several minutes, during which time the user was extremely vulnerable to an enemy attack. In comparison, a trained archer might easily fire a dozen arrows per minute. The range was considerably shorter than contemporary longbows, which might threaten enemy soldiers from 200 meters or more. The weapons were large, heavy, and ungainly, requiring a resting stand for each individual user lest the weight of the barrel pull the user's aim down. The gunpowder needed to fire the weapons did not function in damp conditions, making fighting in the rain almost impossible. Yet, gunpowder weapons also offered some irresistible advantages to the military forces that chose to adopt them, as well as to the leaders of the nations that produced those forces.

Training a musketeer in the use of a firearm could be accomplished within a few hours of practice. Although this would not produce anything approximating a sharpshooter, most early gunpowder armies quickly came to rely upon massed volley fire rather than individual aiming points and emphasized speed of reloading over all other aspects of drill. Using a firearm was much less physically taxing than using a crossbow, which relied upon a significant amount of muscle power, often mechanically assisted, to draw the bow for each round fired. Also, because the production of gunpowder required a significant investment of resources, rulers could control the means of production and hence become the sole suppliers for combat power. This certainly conveyed the advantage that massive amounts of gunpowder were unlikely to simply fall into the hands of peasants, and any rebellion against state authority could be put down or waited out, depending upon the stockpiles of gunpowder available. Finally, and perhaps most importantly, the effect of a wound delivered by a firearm proved much greater than that from an arrow or crossbow bolt. Not only was there a much greater shock effect, as the kinetic force of the projectile transferred to the body of the victim, the shape of the missile also ensured disabling wounds with ragged edges and broken bones. In comparison, sharp arrows penetrated flesh relatively easily but tended to slide around bones rather than shattering them and tended to produce comparatively less blood loss. Another side effect, though one whose cause was not discovered for centuries, was the tendency of bullet wounds to carry pieces of clothing inside the wound, creating almost a guaranteed means of infection given the hygienic practices of the era. Arrows, on the other hand, tended to cut through clothing, creating relatively clean entry wounds with less chances of infection.

Initial European exploration of the Americas revealed a vast land with sizeable indigenous population but very primitive military technology. Most American cultures relied upon natural resources of wood, bone, and stone for their weaponry, although some early metalworking was evident in a few areas. Like other regions of the world, metalworking in the Americas largely started with malleable precious metals. Thus, when Christopher Columbus reached Hispaniola in 1492, he reported that the natives possessed golden trinkets, which they had likely received in trade from larger civilizations on the mainland. The reports of primitive populations with a ready supply of gold and silver spread like wildfire through Europe and a steady stream of would-be conquerors began the trans-Atlantic crossing to the Americas.

When Hernan Cortés landed a force of approximately 600 Spanish explorers on the coast of Mexico three decades later, he had little trouble marching upon the Aztec capital of Tenochtitlan. His warriors, clad in plate metal and armed with steel weapons in addition to their firearms, proved essentially immune to all Aztec weaponry. Fighting between the Spanish and the Aztecs was not combat; it was slaughter. Technically, Cortés was eventually driven from the city for a short period, largely because his forces had been lulled into complacency by their easy conquest and were surprised by an Aztec uprising. However, after regrouping outside the city, Cortés secured alliances with the Aztecs' regional rivals and returned to take the city by storm. In an ironic twist, Cortés ordered his troops to construct trebuchets for the siege of the Aztec capital, as his powder suppliers were running low and he had no capacity to produce more, nor could he expect resupply from abroad. His men must have found the construction order bizarre, as Cortés essentially asked them to technologically regress by a few centuries. Even that regression did not bring them anywhere near the low level of military technology utilized by the Aztecs, which truly demonstrates the massive disparity in technology being utilized and how advanced weaponry can offset a major numerical advantage held by one side. Not surprisingly, none of the Spanish troops had any practical experience constructing medieval siege weapons, even if they thought they understood the principles involved. Upon the first attempted use, the trebuchet flung its projectile straight into the air; when it returned to the ground, it destroyed the siege engine, ending Cortés's foray into constructing ancient weapons. It is likely that any 21st-century soldier asked to construct a flintlock musket would fare little better in his attempts.

With gunpowder's dominance fully demonstrated, European armies commenced efforts to improve both the weapons that employed gunpowder and the defenses to withstand them. At no point, however, was there a significant effort to reverse the gunpowder RMA, or to ignore its effects—it simply became one commonly accepted facet of warfare. By the turn of the 20th century, soldiers carried rifled firearms capable of firing more than a dozen rounds per minute at targets a mile or more distant. Crew-served weapons, such as the water-cooled machine guns of World War I, could fire several hundred rounds per minute, making any infantry advance against prepared positions a futile, suicidal effort. Artillery pieces increased in size, range, and muzzle velocity, while armored vehicles sought to transport troops or engage the enemy with direct fire while providing a modicum of protection to the troops within them. While the form and employment of weaponry gradually evolved, the general concepts of warfare remained the same. Perhaps the final demonstration of gunpowder's primacy is also the simplest—no nation utilizing a military without gunpowder ever managed to defeat a nation using it, regardless of the relative size of the populations and economies involved. While gunpowder did not render its users invincible, it did enable them to conquer the world and create colonial empires that lasted for centuries.

By the 19th century, Western powers had adopted a relatively clear understanding of the acceptable practices of warfare, including which weapons should be prohibited from the battlefield. These "rules and customs" of warfare did not always apply, especially when a Western power engaged in conflict with a non-Western foe. In those circumstances, a total war mentality often emerged, on the presumption that "barbarians" could not be counted upon to adhere to acceptable practices and thus the "civilized" warriors need not limit their own conduct. After all, such limits would only create opportunities for a savage enemy who would perceive them as a form of weakness and immediately move to seize whatever advantage might be gained by such hesitancy. The notion that only "civilized" enemies needed to be respected in regard to the norms of warfare paralleled the earlier European behaviors in the Crusades, when the enemies were considered incapable of following proper military etiquette.

NUCLEAR WEAPONS: THE FIRST STRATEGIC WEAPON

A third RMA of unquestioned importance is the dawn of the nuclear age and the employment of atomic weapons. The Manhattan Project, which resulted in the detonation of the world's first atomic bomb, is an interesting case study of an undeniable RMA. In its case, the American and British governments deliberately set out to engineer a transformation in warfare and to do so while embroiled in the deadliest war in human history. Both nations had ranked among the richest and most powerful nations on earth prior to World War II, although neither nation felt particularly confident about the eventual outcome of the war in early 1942. Warned by leading members of the scientific community that Nazi Germany had commenced work on nuclear weapons several years earlier, President Franklin D. Roosevelt and Prime Minister Winston S. Churchill agreed to combine resources in a crash program to build a superweapon before the Germans could do so.

In the 1920s, Léo Szilárd theorized that it might be possible to split the atoms of certain elements into their component protons, electrons, and neutrons. This could, in turn, trigger a chain reaction of atomic fission, as neutrons of a split atom struck other atoms in turn and continued the process. As each atom split, it would release its particles, along with an enormous burst of energy. If enough atoms engaged in fission at the same time, a massive explosion would be the inevitable result. A lower reaction speed might allow some of the energy to be harnessed, creating a new form of electrical generation capacity. A higher reaction speed, on the other hand, served little purpose for industry or civilian uses—but might allow the creation of the largest and most destructive weapons ever conceived by humanity.

The notion of deliberately inducing atomic fission remained somewhat of a scientific curiosity until the 1930s, when German scientists began to examine the problems associated with putting the theoretical concept into practice. They determined that only certain isotopes would actually function in the required manner. In particular, they focused their efforts upon uranium, an uncommon mineral with radioactive properties. Naturally occurring uranium exists in two isotopes, U-235 and U-238, with the numbers referring to the atomic weight of each isotope. Although U-238 can be used for electricity generation, the isotope needed for an explosive chain reaction, U-235, comprises only 0.72 percent of the average uranium sample. The first major engineering challenge, as a result, was separating the desired isotope from any uranium sample to create a pure enough sample of U-235 for weaponized applications.

German scientists conducted extensive experimentation in their attempts to extract U-235, but their efforts produced only trace amounts, far too little to create even a single atomic bomb. Their efforts were eventually derailed by the German government's demand that scientists focus upon other projects to support the war effort in World War II. U.S. and British scientists commenced their efforts later than the Germans, prompted in part by fears of a German atomic weapon capable of destroying an entire city in a single blast. Famed physicist Albert Einstein, who had fled Germany in 1933 and accepted a research position at Princeton, warned President Roosevelt of the potential dangers of losing the race for atomic weapons, stating, "A single bomb of this type, carried by boat and exploded in a port, [might] very well destroy the whole port together with some of the surrounding territory. However, such bombs might very well prove to be too heavy for transportation by air."[21] Although Einstein did not directly participate in the development project, many of his peers became deeply involved in the quest to refine uranium and turn it into a revolutionary new weapon. British and American scientists agreed to combine their efforts and share the results.

Secure from attack in the continental United States, and with nearly global access to resources, the Allied scientists held major advantages over their German competitors. The U.S. government, which was willing to engage in massive deficit spending as an important means to winning the war, facilitated multiple parallel lines of research into the enrichment process, on the assumption that at least one mechanism was likely to succeed. By early 1945, enough fissionable material had been produced to construct at least three bombs, with enrichment plants working to develop more as soon as possible. The engineers tasked with producing the bomb differed on the optimal design, but once again, the substantial resources available allowed them to try multiple avenues of development. On July 16, 1945, they tested the world's first atomic weapon, at Alamogordo, New Mexico. There is little doubt that the scientific and military observers of the test failed to realize that they had just revolutionized the nature of weaponry. The biggest surprise was how quickly they were able to do so, even allowing for a common purpose and virtually unlimited resources.

When American, British, and allied scientists combined their efforts to construct an atomic weapon during World War II, they did so in a period of immense fear. In 1939, Léo Szilárd posited that the uranium isotope U-235, or the similar element plutonium, might be induced into a massive fission chain reaction, with each individual atom splitting and releasing neutrons, which in turn split other atoms. Each split released a substantial amount of energy in the form of heat and light. If enough material could be compressed to allow the process to rapidly expand, the resulting explosion would equate to thousands of tons of TNT detonating at a single point. The consensus within the scientific community was that such a reaction could be converted into a bomb that might be delivered by an aircraft and which might destroy an entire city in a single moment. On August 2, 1939, Szilárd convinced the eminent physicist Albert Einstein to alert President Roosevelt that German scientists had commenced work upon this type of superweapon. To many of the scientists who joined the Manhattan Project (the code name for the joint atomic effort), the question was not *if* an atomic bomb could be created so much as whether they would be the *first* to construct it. Few thought German chancellor Adolf Hitler would hesitate to use atomic weapons if given control of them.

The nuclear program was at the time the most expensive weapons research program in human history. By the time the first successful test was carried out, on July 16, 1945 at Alamogordo, New Mexico, the U.S. government had already spent US$2 billion on the project.[22] Although Germany surrendered about two months before the test, Japan remained locked in the struggle against the United States. Not only did the Imperial Japanese Army still possess more than two million troops upon foreign soil, the Japanese government had also begun taking the necessary steps to arm the civilian population to resist any Allied invasion of the home islands of Japan. American estimates for the planned invasion of Japan predicted up to two million casualties, with no guarantee that such an invasion would definitely trigger a Japanese surrender. Enemy casualties were likely to be greater by at least an order of magnitude, leaving some to suspect that 20 million Japanese civilians might die over the course of the campaign. Given these grim predictions, it is little surprise that President Harry S. Truman authorized atomic bombings on Hiroshima and Nagasaki. These city-killing strikes not only demonstrated the awesome power of nuclear fission, they also proved the last measure required to secure the Japanese surrender.

The astonishing power of atomic weaponry was not lost upon its witnesses. Clearly, an entirely new capability had been designed, one which could put millions of civilians at risk of instantaneous destruction. Not only did the United States hold a monopoly upon the new superweapon, it also possessed the only aircraft capable of delivering the new devices. B-29 bombers, specially modified to carry atomic bombs, had an effective range of approximately 3,000 miles when carrying nuclear weapons. This effectively put every major population center in the world within range of an atomic strike from an American airbase, to include both enemies and allies.[23] The lesson was not lost upon the Soviet high command or Soviet premier Josef Stalin, who immediately set his scientists upon a crash program to develop atomic weapons and end the American monopoly.

It is no surprise that other powerful nations quickly sought to develop their own atomic programs, rather than leaving the United States as the sole nuclear-armed nation. What is often forgotten, though, is the short period of time in which realistic discussions occurred that considered placing all atomic weapons under an international regime, the newly established International Atomic Energy Agency (IAEA). Many believed that the only means to ensure no nation utilized the awesome destructive power of such devices was to remove them from national control. If implemented, this plan might have not only prohibited nuclear exchanges but also limited the capability of nations to go to war using conventional weaponry, out of fear of provoking an atomic strike. The most optimistic leaders saw the IAEA as an agency akin to the United Nations, meaning an international body that might prevent the need to engage in warfare to settle international disputes. Unfortunately, the victorious Allies of World War II, who had cooperated and set aside political differences in the face of a common enemy, proved far less amenable to working together once the threat had dissipated. The notion of placing atomic weapons under international control, and banning their development by individual states, never got enough traction to occur.

For over three years, the United States maintained its nuclear monopoly while relations with its erstwhile ally, the Soviet Union, rapidly disintegrated. The future of nuclear power was a major point of contention among the victorious Allies. Soviet premier Josef Stalin was strongly resentful of not being informed of the Manhattan Project prior to the attacks on Japan. He feared that such weapons might be turned against his nation—a fear that was stoked by public statements made by top military and civil leaders suggesting attacks against the Soviet Union while it remained weakened by the war. To prevent or at least complicate such attacks, Stalin's forces imposed a harsh occupation throughout Eastern Europe and East Asia, essentially creating a massive land buffer between his homeland and a newly hostile West. Soviet air defenses across this buffer zone might shoot down atomic bombers long before they reached Moscow or other critical targets. Soviet agents instigated revolts within several vulnerable states, seeking to expand the territory under Soviet domination and hence improve its security. Soviet direct aid to the communists in China sought to tip the balance of the Chinese Civil War in their favor, also in anticipation of a friendly regime that might forestall any attacks from that direction.

Having seen the devastation wreaked by two atomic bombs, Stalin also ordered a crash program to build a Soviet atomic device and return to military parity with the United States. Resource costs proved irrelevant; only success in the program mattered. On August 29, 1949, the first Soviet atomic bomb exploded in Kazakhstan, ending the American monopoly and triggering a new arms race. The explosion also ended any hope of placing atomic weapons under international control. Although the United States assured its fellow democracies that it would place them under the American umbrella of nuclear protection, both the British and French governments chose to develop their own atomic arsenals, assuming that the United States would not risk its own nuclear annihilation by launching atomic strikes on behalf of its allies. Each nation's program succeeded, with the first British detonation in 1952 and the first French atomic explosion in 1960.

In 1964, the People's Republic of China followed suit, although on a much smaller scale. In time, India (1974), Pakistan (1998), and North Korea (2006) openly demonstrated their nuclear capabilities. Many consider Israel a nuclear state, although the Israeli government refuses to confirm or deny its possession of nuclear weapons, considering a deliberately ambiguous policy to be a valuable deterrent against possible aggression from its neighbors. In each case, fear of a rival state proved a significant driving factor in a nation's decision to invest the time and resources necessary to develop nuclear weapons. The Soviets pursued atomic weaponry out of fear that the United States could not be trusted with a nuclear monopoly. Britain and France both decided to develop weapons to secure their own positions in Western Europe against possible Soviet aggression, in part because the political leaders of each state did not fully trust the United States to risk a nuclear war on behalf of an ally. The People's Republic of China began its weapons program in the aftermath of the Korean War, having been threatened with nuclear attack by General Douglas MacArthur. India, which had fought wars against China for centuries, could not tolerate a nuclear unbalance in its region. In turn, the successful Indian nuclear weapons program guaranteed that Pakistan would pursue nuclear weapons to offset its most dangerous enemy. North Korea, often called the "Hermit Kingdom" for its extreme isolation from the rest of the world, commenced a nuclear weapons program in part to guarantee the continuation of the Kim regime.

When the first atomic bombs were used in warfare, their development required shipment from New Mexico to Tinian Island in the Marianas. Most of the transportation was provided by USS *Indianapolis*, under a cloak of the highest secrecy. The mission was so classified that when the *Indianapolis* was attacked by a Japanese submarine after delivering her cargo, it took more than three days for her to be missed. On Tinian, each bomb was loaded upon a specially modified B-29 bomber for the 3,000-mile flight to its final target. In the aftermath of the war, though, with Axis territory partially occupied by U.S. troops and a series of forward airbases literally ringing Soviet territory, the U.S. president might have ordered a strike that would descend upon Moscow in a matter of a few hours. With the advent of the intercontinental ballistic missile (ICBM), the time from order to detonation of a nuclear device over the enemy capital dropped even further, to less than an hour under optimal conditions. Advanced medium-range ballistic missiles in Turkey (or those situated by the Soviets in Cuba in 1962, for that matter) could theoretically decapitate the enemy government in just a few minutes after launch. With this "sword of Damocles" looming over both the United States and the Soviet Union, the possibility existed that even a minor miscalculation might lead to catastrophe—and with the far shorter time periods from planning to launch to detonation, the opportunity to defuse a crisis all but disappeared. Thus, the previous assumptions regarding whether or not to launch an attack, on what scale, and with what objectives became radically, permanently altered.

Like gunpowder weapons, nuclear armaments have undergone qualitative improvements in their design and delivery since their inception. The first atomic bombs required either very heavy bombers or delivery by ships or ground vehicles. By the 1950s, however, it had become clear that long-range missiles might offer a more effective delivery method, particularly if they could follow ballistic paths that effectively could not be intercepted. ICBMs could theoretically travel to anywhere on the globe in under an hour and would leave the earth's atmosphere en route, allowing them to attack targets on an almost perfectly vertical angle and a terminal velocity far too rapid to be shot down. Multiple-reentry warheads allowed a single missile body to launch multiple attacks toward widely dispersed targets, making nuclear attacks both more efficient and more effective.

On November 1, 1952, American scientists detonated the first hydrogen bomb. This device, which exploded with the force of 10 megatons of TNT, was 500 times more powerful than the bomb that destroyed Nagasaki. It required an atomic bomb to serve merely as a trigger, the source of energy needed to initiate a hydrogen fusion reaction. This fusion reaction, in which two atoms of hydrogen combine to form a single atom of helium, is the same process by which the sun produces its energy—and constituted a massive increase in the potential damage that could be done by a single warhead. Just three years later, the Soviets detonated their first hydrogen bomb, shocking the West by how quickly they had seemingly closed the technological gap.

American and Soviet designers followed different assumptions in their pursuit of ever-deadlier nuclear weaponry. In general, American guidance systems tended to be far more effective, meaning an ICBM launched from the United States would hit in a much smaller target area. This also allowed a higher number of smaller warheads to be fitted on a single missile for the same net effect. Of course, hitting a target the size of a city is not a particularly difficult proposition, relatively speaking—which led many to question why the United States had focused so heavily upon precision guidance. Hitting a target hardened against nuclear attack, on the other hand, required much greater accuracy. While it is possible to create passive defenses capable of withstanding a nuclear detonation in their vicinity, it is almost impossible to harden a target sufficiently to withstand a direct hit. This increased precision was an outgrowth of American strategic assumptions about nuclear warfare (discussed in subsequent text) and opened new strategic options to war planners in the United States.

The Soviet Union did not have the same level of precision machine tools as those used by the United States, and its electronics industry was also not as advanced. Thus, rather than essentially chasing every American advance and remaining doomed to a perpetual second place in the nuclear arms race, Soviet planners simply went in an entirely different direction. Instead of accuracy, Soviet weapons designers primarily focused upon explosive yield. As a result, the largest manmade detonation in history occurred during a Soviet weapons test. The Tsar Bomba, detonated on October 30, 1961, released an explosion the equal of an incredible 57 megatons of TNT, more than 10 times higher than the energy of all of the weapons used during World War II combined, including the atomic weapons used against Hiroshima and Nagasaki. Such a blast could reduce any city on earth into a smoking ruin in an instant. Although far too large for easy delivery when it was created, the continual miniaturization process made it entirely possible that a comparable-yield warhead might be mounted upon a missile. Despite American beliefs that the Soviet Union was an aggressively expansionist state that would not hesitate to start World War III any time it expected to win, the entire Soviet arsenal was essentially designed to serve as a retaliatory force—it did not have any possibility of destroying U.S. missiles in a first strike.

One of the hallmarks of a true RMA is that the rules and assumptions of previous military behavior do not function under the new paradigm. The same was certainly true after the advent of nuclear weapons. Any attempt to simply slide nuclear weapons into existing national strategies or war plans would have been extremely inefficient, at best, and almost certainly guaranteed to inflict enormous and unacceptable levels of civilian casualties. Employing nuclear weapons without causing outrageous collateral damage was almost impossible. If one resolved to employ only atomic weapons against fielded armies, the enemy would likely maintain troops only in populated areas, or disperse the troops so thoroughly as to make atomic strikes ineffective. If one vowed to use them only in a retaliatory capacity, it not only ceded the initiative to the enemy, it also provided an insurmountable advantage to the side with the largest conventional forces. As is almost always the case with an RMA, the innovation far outpaced the ability to devise an effective means of using it.

In 1959, Bernard Brodie published a seminal classic of nuclear strategy. His work *Strategy in the Missile Age* argued that deterrence would be best achieved by possessing a demonstrated second-strike capability, even in the face of a surprise nuclear attack. Of course, in some ways, he was essentially summing up what had been American practice for more than a decade. But in other ways, he broke new ground by pointing out that ICBM silos could be hardened to withstand anything but a direct hit from a nuclear missile.[24] In 1960, Herman Kahn drove the discussion of how nuclear weapons might be employed in a radical new direction. In *On Thermonuclear War*, Kahn made the case that nuclear weapons might possibly be employed on a limited basis, in contradiction to the standing assumptions of many political and military leaders. In his estimation, the use of one or even a handful of nuclear weapons did not inevitably guarantee a full exchange of nuclear attacks, so long as the initial usage remained confined to a relatively limited area. His argument included the development of an escalation ladder that laid out dozens of types of war, in increasing levels of seriousness and probable damage. Interestingly, he placed the first (and least destructive) use of nuclear weapons on only the 15th step of his escalation ladder.[25]

The notion that nuclear deterrence might not be an all-or-nothing proposition, and that even nuclear warfare might be conducted on a limited basis, contradicted the dominant perspective of the Truman and Eisenhower administrations, both of which tended to consider the use of nuclear weapons as a threshold that, once crossed, could not be halted short of the complete destruction of one or both sides. In fact, Eisenhower's threat of massive retaliation through nuclear attack in response to any Soviet aggression rested upon the idea that the United States was willing to escalate to its maximum capability at a moment's notice.

To truly understand the concept of a maximum escalation situation, consider the behavior of both the United States and the Soviet Union during World War II. In 1939, the Soviet Red Army was by far the largest army in the world, with approximately five million troops in uniform (drawn from a total Soviet population of only 130 million). Yet, it was a hollow shell—its best leaders had been purged by Josef Stalin's commissars, largely due to fears that they might prove politically unreliable in a crisis. On June 22, 1941, the crisis erupted, in the form of a massive German invasion of nearly 150 divisions on three axes of advance. In the first six months of combat, the Red Army lost 5 million troops, as well as 20,500 tanks and 21,000 combat aircraft.[26] It took over two years of maximum effort to return the Red Army to its prewar size and almost three years to recapture the territory lost in the initial months of fighting. Over the course of four years, the Soviet Union managed to enlist over 34 million troops, waging an all-out effort for the survival of the state.

The U.S. Army at the end of 1940 stood at only 269,023 troops, despite the fact that Europe had been at war for more than a year. Prewar preparations left the force at 1,462,315 in 1941, three weeks after the Japanese struck Pearl Harbor. More than two years elapsed, during which time the American homeland was essentially immune from attack beyond submarines targeting coastal shipping, before the United States could mobilize sufficient forces for anything larger than local offensives. For D-Day, almost three years after the German invasion of the Soviet Union, the United States committed only seven divisions, including two divisions of paratroopers dropped from the skies over Normandy. Even at the time of the German surrender, on May 8, 1945, the U.S. Army stood at only 61 divisions in Europe, despite its best efforts to induct, train, and transport forces to support the Germany First strategy. In comparison, the Soviets had over 600 divisions of ground forces. The United States made up much of the disparity through the development of an enormous air arm, comprising nearly 2.4 million personnel flying more than 100,000 combat aircraft.

Despite the incredible devastation that each side in the Cold War could inflict upon its rival, the concept of mutually assured destruction pushed each side into a certain degree of rational, predictable behavior. Neither the United States nor the Soviet Union was willing to engage in a nuclear exchange as long as the initiator stood to take as much damage as its enemy. However, in the 1970s, American scientists began to seriously consider the question of whether or not it might be possible to establish an antiballistic missile defense system. The 1972 Antiballistic Missile Treaty limited the number of sites that the United States and the Soviet Union could establish for such a system and required either signatory to provide a six-month notice of any intention to abrogate the treaty. In the 1980s, President Ronald Reagan commenced a serious push to build just such a system. Historians are divided on whether Reagan really thought such a system would work or whether he was engaged in bluffing the Soviets to convince them to sign a series of nuclear arms limitations treaties. Either way, research commenced into a variety of potential systems. In 2002, President George W. Bush formally withdrew the United States from the Antiballistic Missile Treaty, ostensibly to set up a number of limited sites to defend the United States against a "rogue nation" such as North Korea or Iran. Public tests of the various systems showed a mixed track record, which was hardly reassuring for members of the public, who might be on the receiving end of a rogue strike. The only portion of the system that seemed to work almost flawlessly was a series of observation satellites capable of detecting a missile launch, providing at least a small amount of warning for any inbound strike.

It is illustrative that would-be nuclear powers in the late 20th and early 21st centuries have struggled to achieve what more technologically and economically developed nations achieved decades earlier. North Korea conducted an atomic test with great fanfare in 2006, but failed to impress the world when its plutonium device yielded less than one kiloton of explosive power. Iran, which as of this writing continues to deny any ambitions to develop nuclear weapons, has struggled to refine its uranium stockpiles to weapons-grade enrichment levels. Protests that the Iranian regime intends to use enriched uranium for peaceful purposes have largely been ignored, primarily because Iranian nuclear scientists did not halt their enrichment efforts when they reached the necessary purity required to sustain a fission reaction capable of generating electrical power.[27]

ROBERT PAPE AND AERIAL PUNISHMENT

In 1996, Robert Pape published *Bombing to Win*, an examination of the efficacy of aerial bombardment campaigns throughout history. He separated these campaigns into two primary divisions: punishment and denial. To Pape, punishment constituted an effort to inflict sufficient damage upon an enemy nation as might be necessary to coerce it into accepting an attacker's demands. Classical airpower theorists such as Italy's Giulio Douhet and Britain's Hugh Trenchard had envisioned just such a campaign. Each assumed that a civilian population subjected to the horrors of aerial bombardment would essentially revolt against its own government and compel the nation's surrender to end the punishment campaign. Douhet assumed that airplanes would drop high explosives and chemical weapons in an indiscriminate matter, killing thousands in a single raid and demonstrating the futility of resistance. Trenchard did not believe so much physical damage would be needed, as in his mind, the psychological effect of seeing bombers overhead would suffice to terrify the population into compliance.[28]

Pape further divided punishment campaigns into two forms and examined historical examples of each to devise a broad theory of airpower as an instrument of national power. He found that punishment campaigns almost never work—rather than turning the population against the government, they tended to drive the two closer together. Douhet and Trenchard had both mistakenly assumed civilian victims of bombing would blame governments for not performing the most basic function of defending the population from attack. Instead, the victims tended to blame the enemy nation directly responsible for their misery—and paradoxically redoubled their efforts to support the armed forces tasked with prosecuting the war. Interestingly, Pape observed that his theory of punishment campaigns could be accurately applied only to conventional warfare—the awesome power of even a limited nuclear strike might serve to coerce an enemy government, or failing that, an enemy population, to surrender as a means of avoiding further punishment. To Pape, even the most unlikely of nuclear warfare scenarios could not be dismissed by a government or military facing a nuclear-armed foe, as the costs of a nuclear miscalculation were too high.[29]

Pape's fundamental purpose was to argue that conventional air power has been most effectively used in what he termed a denial strategy. In this form of utilization, aerial attack, or the threat of it, can deter an aggressor from engaging in warfare—but it will not drive an army out of ground it occupies once that territory has been secured. Aerial attack can eliminate the most advantageous routes of march, by destroying bridges, rail lines, and roads, but it must do so before the enemy army reaches them, rather than once it is in their rear areas, even if it serves as a key line of supply. Almost by accident, though, Pape demonstrated the key concept of RMAs—that they do not follow the norms, standards, assumptions, theories, or doctrines of pre-RMA warfare. Thus, an RMA does not simply change the mode or method of human conflict; it renders it obsolete and creates an entirely new concept in its place.

THE ROLE OF REVOLUTIONS

The phalanx, gunpowder, and nuclear weapons all represent examples when a major transformation of warfare occurred, over a significant period of time. With the exception of atomic weapons, the fact that an RMA was under way was probably not evident to contemporary observers—but the changes continued regardless. While it is difficult to pinpoint the exact moment when the changeover occurred (again, excepting nuclear weapons, when the said point is fairly obvious), there is no question that societies that adopted new technologies or systems gained an immutable advantage over those that did not.

Are military robots an RMA? They certainly offer an enormous potential advantage to early adopters, especially if they are used against an opponent without the same technology available. Ethicist Laurie Calhoun certainly sees them as a fundamentally different form of war making, noting, "With the advent of weaponized drones, it has become possible to kill without sacrificing troops, but the focus on lethality remains the same."[30] Ann Rogers and John Hill state the same concept, but in even more stark terms, when they argue, "After centuries of technological innovation, militaries have finally deployed weapons that routinely remove human risk entirely from one side of the equation while expanding it exponentially at the other."[31] Of course, the utility of military robots lies primarily in the ability to keep one's own forces out of harm's way—but does that increase the likelihood of using force as a tool of national policy? The next chapter will provide an overview of the development of the robotics industry and how it has been utilized to drive human productivity, explore previously unreachable locations, and push the limits of military activities. This context will then be used as a mechanism to drive the discussion of current and near-future employment of military robots.

CHAPTER 3

The Long Gray Line

An improvement of weapons is due to the energy of one or two men, while changes in tactics have to overcome the inertia of a conservative class; but it is a great evil. It can be remedied only by a candid recognition of each change, by careful study of the powers and limitations of the new ship or weapon, and by consequent adaptation of the method of using it to the qualities it possesses, which will constitute its tactics. History shows that it is vain to hope that military men generally will be at the pains to do this, but the one who does will go into battle with a great advantage, a lesson in itself of no mean value.

—Alfred Thayer Mahan, 1890

There is a long history of humans envisioning machines to take their place on the battlefield. Well before the procurement and fielding of military robotics, storytellers and strategists imagined how mechanical troops might conduct warfare. Those visions have served to influence the actual development of military robotics, as designers seek to fill the niches already established by imaginative writers. At the same time, other engineers have designed military robots without reference to science-fiction movies or tall tales of automata warriors, resulting in a wide variety of configurations for robots that perform an almost limitless number of combat tasks in the modern world.

For thousands of years, thinkers have envisioned ways to substitute for human beings on the battlefield. Such ideas have included magical creatures, military vehicles, and artificial combatants. While their imaginations might have invented an almost limitless number of ways to inflict damage upon the enemy without risking one's own forces, the practical applications of such ideas proved far more limited prior to the 20th century. Given the costs of warfare, particularly when conducted on a large scale, it is only natural that people would seek any alternative means available to undertake violence with impunity, attacking the enemy while remaining immune to counterattack.

ANCIENT DEPICTIONS OF ARTIFICIAL WARRIORS

Classical literature is replete with examples of autonomous machines capable of engaging in combat operations. While the means of animation varied over time and by the source material, the concept of a powerful humanoid capable of independent action and reaction with its environment certainly resembles the modern conception of what military robots might be in the near future, particularly if they are designed to interact with and work alongside human counterparts.

Ancient Egyptian pharaohs devoted enormous sums from the state treasury to build their own tombs, and much of our modern understanding of their society comes from the relics left behind inside these enormous structures. Egyptian rulers believed in an afterlife where they would continue to be served by the people, animals, and objects that they took with them into their tomb. In addition to servants killed and placed in the tombs to accompany their leaders on the journey to the afterlife, there were also often physical representations of living creatures, in the form of statuary, that might have been expected to serve a far-more-than-decorative function in the afterlife. Included in these statues are numerous figures that might serve in a combat function, including humans, animals, and some seemingly hybrid creations. While not robots by any stretch of the imagination, these figures certainly demonstrated that the Egyptians could envision the possibility of a nonliving servant that might perform functions and tasks previously handled solely by humans or living beasts.

In approximately 800 BCE, the first written versions of the Homeric epics began to appear. Homer's best known works, *The Iliad* and *The Odyssey*, both purported to chronicle the behavior of Greek heroes from the Mycenaean Era, four centuries earlier. In *The Iliad*, the Myrmidons, followers of Achilles, are often described as something other than human, although modern translations have marginalized or ignored entirely their depictions as automata. While this might indicate a dramatic flair for describing the machinelike precision of these highly trained warriors, or differences in translations that have emerged over thousands of years of reprints, it is also conceivable that the epic is describing a mythical fighting machine. The Greeks were comfortable with the concept of automata, many examples of which have been discovered in archaeological excavations, although the primitive automata found in the city-states have largely been garden water features, not free-moving warriors.

Ancient Chinese rulers focused a great deal upon their expectations of the afterlife and the challenges they might face after death. Like the Egyptians, they believed in the need to stock provisions and luxuries for their continued existence after death. The tomb of Emperor Qin Shi Huang included more than 8,000 terra-cotta sculptures of warriors, a massive army of nonliving statues that might have served only a decorative purpose or been considered a protective force that would guard and shepherd the emperor forward into the unknown. Such representations might have served a solely religious or mythical function, but it requires little speculation to assume that the creators of these statues imagined them as fully functional, moving warriors if they could only be animated. Once again, these figures definitively show the human capacity for imagination regarding the use of artificial substitutes for the battles that might be faced in another plane of existence.

There is little evidence from the classical period that any degree of automation actually permeated military forces. However, the notion of automata certainly existed in other aspects of civilization. Engineers of the classical world harnessed hydraulic power as a means to create automatic amusements, most notably in the famed Hanging Gardens of Babylon, where visitors could enjoy an entire automatic theater powered entirely by water pumped through the machines. Similar automata, which served little if any function beyond amusement, appeared throughout the Roman Empire and played a prominent role in the expansion of Arabic culture.[1]

Some early inventors had invested considerable time and expense trying to emulate life with automata. Often, their greatest contributions to technological innovation came not in the final products of their imagination but in the precursor technologies needed to make them functional. For example, Greek inventor Archytus created a piece of automata shaped like a dove around 400 BCE, which, while interesting, is not nearly so important as the fact that his wooden dove was powered by compressed air rather than water. Such a power source required much greater precision in the production of individual moving parts and also a means to compress, store, and release air in a controlled manner. Midway through the 9th century, the Banu Musa brothers published *The Book of Ingenious Devices* in Baghdad. This Arabic text not only consolidated many of the earlier designs of automata into a single volume, it also demonstrated a massive shift in the geographic center of innovation to the Islamic world.[2] Nearly four centuries later, Abu al-Jazari's *The Book of Knowledge of Ingenious Mechanical Devices* provided detailed instructions for the creation of dozens of machines with practical uses, including automata. It showed a clear maturation of the ideas first published centuries earlier and recognized that automata did not have to be devised solely for novelty value. Unlike previous creations, al-Jazari's devices emphasized the possibility of replacing human labor through mechanical creations, demonstrating a significant leap forward in the use of automata. Internal schisms within the burgeoning caliphate, coupled with the immense costs of fabricating al-Jazari's creations, undoubtedly hindered their widespread adoption.[3]

Leonardo da Vinci, widely considered one of the greatest inventors of all time, demonstrated more than a passing fancy with automata, although he is rightfully better known for his artistic work and his military machines. In 1497, da Vinci designed an "automated knight" that could move and possibly emulate limited speech, according to contemporary viewers. Although the knight probably had no military application, it certainly inspired those who saw it to imagine an army of automated warriors. According to the surviving designs, the knight relied upon steam power and pulleys for its movements.[4] The extant blueprints would not produce a functional machine, but da Vinci was known to deliberately include flaws in the written versions of his designs to prevent theft by his competitors. In 1515, da Vinci presented an automated lion to King Francis I of France, a transparent and successful attempt to maintain the king's patronage for his other projects.

In the early 18th century, Jacques de Vaucanson created a series of celebrated automata that became the delight of the French nobility. Two—the flute player and the tambourine player—were humanoid automata that played real instruments, with a variety of songs available to the user. To increase the already-disturbing realism of these creations, de Vaucanson covered both in real skin. Perhaps his most famous creation, the digesting duck, appeared to be a mechanical bird that could walk, quack, eat grain, and even defecate. It was later proven that the latter function was actually simulated by the release of predigested grain in a hidden pouch, rather than any actual consumption of corn. Nevertheless, not only did his devices stimulate the popular imagination, they also required the creation of the world's first rubber tubes, a key innovation that soon began to appear in a multitude of other devices.[5]

Two decades after de Vaucanson's duck, Friederich von Knauss built a pair of automata, one capable of writing short phrases and the other capable of performing short musical tunes. Their sophistication and lifelike appearances amazed the aristocracy of Europe but were soon in turn outmatched by the efforts of Swiss watchmaker Pierre Jacquet-Droz. His automata, named the Writer, the Musician, and the Draughtsman, incorporated a rudimentary programming system that enabled a variation in their behaviors. In some ways, Jacquet-Droz had created the first primitive computer, solely for the purpose of enabling his mechanical humanoids. Perhaps if Jacquet-Droz had continued to replicate common professions, he would have eventually designed the "Soldier," although like his other creations, it would have been a simulacrum, not a functional, mobile robot.[6]

At the turn of the 19th century, Henri Maillardet built automatons capable of writing and drawing, using a cam-based memory system to allow programming. This also enabled Maillardet's creations to give a convincing performance of interacting with their environment, even though they merely followed a programmed sequence of events. In many ways, his automata were far less complex than a mechanical loom of the same period, but they certainly evoked a much stronger response from the citizenry of Europe.

   The period of revolutionary wars in Europe largely cooled the ardor of the populace for automata, as most innovators turned to more mundane innovations of military utility. Also, mechanical automata had essentially reached the limits of what could be accomplished via the available power sources and without substantial improvements in programming and calculating technology. Such innovations, however, were certainly in the works, as evidenced by the efforts of a number of innovators. In 1805, for example, Joseph Marie Charles Jacquard invented the punch card system, a form of programming to transform the function of mechanical looms. Jacquard's approach allowed the creation of incredibly complex woven patterns that, once created and input on a card, no longer required the intervention of an expert weaver on the loom. Not only did this serve to replace one of the highest-paid trade specialists within the textile industry, it also increased the speed with which such patterns could be replicated on the high-speed looms.

   British mathematician and inventor Charles Babbage sought mechanical means to conduct mathematical calculations. In 1822, he designed the difference engine, a massive calculator capable of not only performing basic arithmetic but also using mechanical means to approximate logarithmic and trigonometric functions. Unfortunately, while the design was sound, the market for such a machine did not exist that would have justified the massive costs required for its construction.[7] Despite the disappointment that he could not get a benefactor to advance the funds needed for his creation, Babbage continued to advance his designs, and, in 1837, created plans for the analytical engine, an even larger project that can be considered the forerunner of the modern general purpose computer. Not surprisingly, this device was also not constructed, and plans for its construction essentially disappeared into obscurity for more than three decades.[8] Babbage's machines might have benefited from the system of machine logic invented by George Boole in 1854 and presented in *The Laws of Thought*. Boole's concept, often referred to as "Boolean logic," attempted to convert abstract thought into a concrete, digital form. This notion underpins virtually every commonly used computer programming language in existence today and is therefore a key component of all military robotic systems.

EARLY INDUSTRIAL ROBOTIC SYSTEMS

   The earliest precursors of military robotics began to appear at the turn of the 20th century. Nikola Tesla demonstrated a radio-controlled boat for the U.S. Navy in 1897, on the assumption that naval officials would quickly recognize the feasibility and potential tactical utility of a radio-controlled torpedo. Like many inventors of the era, Tesla's belief in the rationality and enthusiasm of the navy proved incorrect, although a number of other naval forces in the world quickly grasped the potential represented by a torpedo that could be steered from a shore installation or from aboard a ship.[9] At the time, existing torpedoes depended upon either compressed gas or a chemical reaction for their propulsion and could not be steered once they were fired upon a predetermined path. Experiments with wire-controlled torpedoes had proven relatively unsuccessful, as they allowed a certain degree of guidance but also created an unacceptable level of drag in the water and the need to unspool the guide wires severely limited the range of the weapons. Tesla's improvement upon these models, allowing the path of the torpedo to be adjusted via wireless signals, represented a major leap forward in remotely controlled weaponry and offered a glimpse into the future of command-detonated explosives and remotely controlled vehicles, but it was simply too advanced and expensive for the U.S. Navy, which far preferred to rely upon geography as the primary means of American coastal defense. By the time Tesla demonstrated his weapon, the prevailing conventional wisdom was that the United States, protected by the Atlantic and Pacific oceans, was simply too distant for any aggressor to reach in the era of steam-propelled naval vessels that relied upon coaling stations for any significant power projection. Thus, it made little sense to spend money upgrading coastal fortifications that would never be tested by enemy action, when the money could be spent upon improving the expeditionary fleet instead. Of course, the fact that the torpedo might prove deadly in naval engagements provoked little interest, either, on the assumption that all future naval battles would be fought at battleship ranges, far in excess of any torpedo in existence. Despite the experiments in submarine warfare that had been carried out over the preceding century, it was simply inconceivable that any ship relying upon torpedoes as a primary armament might offer a significant threat to the vaunted American navy.

Similar skepticism permeated both the U.S. Army and the U.S. Navy when it came to the incorporation of aircraft into the military inventory. Many leaders considered them simply a novelty or an unreliable means of scouting terrain ahead of an advance rather than a weapon of war in their own right. When the Wright brothers sought to demonstrate the value of their heavier-than-air powered flying machines, they encountered almost no interest in the United States. As such, the young inventors were forced to search for markets abroad and established their first flying school near Paris. Other nations at least reasoned that aircraft might be worth a small investment and began to incorporate the new machines into their inventories. Once the usefulness of the airplane had been established, its numbers quickly swelled in the most powerful armies of the earth, while for the U.S. Army, it largely remained a novelty.

During World War I, the airplane truly came into its own as a vehicle of warfare. Thousands of airplanes flew over the battlefields of Europe, first as observers and artillery spotters but soon as weapons with offensive power. Not surprisingly, airplanes grew in size and capability during the war, with model versions becoming outdated in 6 months and completely obsolete in 12. The U.S. resistance to incorporating aerial forces came back to bite the American Expeditionary Force when it was committed to the war in 1917. While the U.S. Army managed to field several hundred pilots, they were forced to fly French and British aircraft, often used and somewhat out-of-date models, due to the poor production capability of American manufacturers.

Airplanes offered certain potential advantages that might have influenced the stalemate in the trench lines of the Western front. First, of course, they could fly over the lines and attack rear areas that were more likely to be unprotected and hence vulnerable to direct attack. As airplanes' speed and range increased, it became feasible to launch attacks upon enemy cities, rather than solely military targets, although the lift capacity of aircraft made such attacks relatively minor. Second, aircraft could attack ground formations at a much greater angle than any other instrument of combat. This largely negated the protections of the trenches, which served as an excellent form of cover against the machine guns and artillery pieces of the era but which tended to be open to the sky. Third, bombs dropped from airplanes, while incredibly imprecise, had both a larger volume of explosives relative to the weight of the bomb than a comparable artillery shell and much greater penetrating power due to the kinetic effects of gravity. Thus, airplanes might potentially destroy fortifications that had proven immune to artillery barrages, if the pilots could manage to place their bombs on target.

Charles Kettering, a research engineer with General Motors, sought to combine many of the most attractive elements of aircraft and artillery barrages. He invented the world's first aerial torpedo, essentially an airplane that could be programmed to turn off its engine after a set distance. At that point, the device turned into a very inefficient glider, theoretically slamming into enemy defenses at a high angle of attack and blasting holes in their positions.[10] His invention, dubbed the Kettering Bug, used a control system designed by Elmer Sperry, founder of the Sperry Corporation. Sperry had designed the first gyroscopic autopilot in 1912 and demonstrated it at a Paris air show in 1914.[11] In testing, the Kettering Bug proved relatively inaccurate. Although it had a range of up to 50 miles, it could be counted upon to strike only within a few miles of the target rather than the precision attacks that Kettering envisioned. Not surprisingly, the U.S. military chose not to purchase and deploy any of the Kettering Bugs, and although some Allied leaders showed a modicum of interest, none of them was sent to Europe prior to the armistice.[12]

THE ROBOTICS OF WORLD WAR II

Although Tesla's and Kettering's inventions received little interest in the United States, some of their overarching ideas certainly caught the attention of weapons designers elsewhere in the world. In the Soviet Union, Tesla's wireless torpedo led to the idea that a much more complex weapon might be controlled using the same principle. By 1939, Soviet engineers had adapted the concept to develop the "teletank," a remotely operated armored vehicle with a completely automated self-reloading gun. Soviet commanders hoped that teletanks could be used to blast holes in enemy positions while allowing their human controllers to remain safely ensconced in a command tank beyond the range of enemy guns. In practice, it was a terrible waste of resources, as there was little way for the operators to actually see what was within range of the teletank guns and the command tanks often had to fire their own rounds to destroy the teletanks from behind, lest they be captured intact by an enemy and turned against their Soviet masters. The only operational deployment of the teletanks came in the winter war with Finland in 1939–1940; in fact, had the Finns captured any teletanks, they likely would have quickly realized the terrible design flaws and absolutely refused to have anything to do with them. The final, and probably most successful, use of teletanks came during the initial German assault upon the Soviet Union in 1941. During that desperate period, the teletanks were deployed essentially as decoys, forcing the Germans to expend at least a few rounds of antitank ammunition to destroy an otherwise completely worthless piece of gear.[13]

For their part, the Germans were not immune to the siren song of remotely operated weapons, both of the ground and the aerial variety. As part of their interwar buildup, German engineers created the remotely operated Goliath, a 200-pound tracked vehicle operated via a control wire and an electronic command box. The Goliath carried a large explosive charge and was designed to be driven underneath enemy tanks and detonated. In practice, while each one was not as expensive as a teletank, its overall usefulness was about the same. The command wire was extremely vulnerable (and only 500 meters long), and without it, the Goliath could neither be moved nor be detonated. It was agonizingly slow and required the operator to remain within visual range, which often meant the operators had to expose themselves to enemy fire in order to slowly advance their charges. Although several thousand were manufactured, there is little evidence that any of them were ever used in World War II for their primary purpose: the destruction of enemy tanks. Hundreds of them were eventually employed later in the war as command detonated mines and often emplaced on bridges that needed to be destroyed at the last possible moment in the face of an Allied advance.[14]

One of the more feared German weapons of the war expanded upon the idea of the Kettering Bug. The *Vergeltungswaffe 1* (Vengeance Weapon 1), more commonly known as the V-1 flying bomb, used a jet engine and an explosive load to attack Allied urban centers. Like the Kettering Bug, the V-1 flew on a preprogrammed route, although it did so much faster and farther than its World War I ancestor. At first, the V-1s sowed enormous fear with their unpredictable attacks and the buzzing sound they made en route to the target. However, the British soon realized that the V-1s flew on a straight and level path and quickly devised tactics to defeat them in flight. When a flight of V-1s was detected, interceptor planes chased them down, flew a parallel course, and then literally bumped their wings to knock them off-balance. Because the V-1s had no means to realign themselves, or even to detect the unbalancing activity, when the wings were tipped far enough off level, the flying bombs would simply drive themselves into the North Sea. Even with all of their flaws, the V-1s still offered a mechanism to strike at the enemy civilian populations and the 8,000 launched in the war killed at least 6,000 civilians.[15] Further, the threat of increased V-1 attacks caused a shift in Allied strategic priorities, forcing a deliberate drive to capture and destroy the launching zones along the Dutch and Belgian coastal areas to eliminate the strikes, when the previous strategy had called for an invasion of Germany as quickly as possible.

The most well-documented American foray into the realm of remotely operated aerial vehicles during World War II was Project Aphrodite. This ill-conceived program called for packing obsolete bomber aircraft with as many explosives as possible and just enough fuel for a one-way trip to an enemy target. The aircraft had a remote-controlling apparatus that allowed it to be flown via wireless controls from a master aircraft and could be turned into an enormous cruise missile that would be forced into a suicidal dive upon an enemy position. While allowing for much greater accuracy than the V-1 or the Kettering Bug, there was an enormous flaw in the design of the Aphrodite system. The soon-to-be-destroyed aircraft required a pair of pilots for the takeoff phase of its flight. These pilots remained with the aircraft until it reached a safe altitude and then turned on the remote control system and bailed out of the airplane through the emergency escape hatches. Not surprisingly, Project Aphrodite relied entirely upon volunteer crews for this incredibly hazardous duty.[16]

Project Aphrodite launched only 26 attacks at 15 targets and might have simply faded into obscurity were it not for the identity of one of the volunteer pilots.[17] On August 12, 1944, Lieutenant Joseph P. Kennedy, Jr., took off in a Project Aphrodite airplane with his copilot, Lieutenant Wilford J. Wiley. Their aircraft was designated to attack Mimoyecques, a massive fortress near Pas-de-Calais. After only a few minutes of flight, the aircraft suddenly exploded, instantly killing Kennedy and Wiley.[18] Biographers of the Kennedy family have suggested that had he survived, Joseph P. Kennedy, Jr., would have run for the presidency rather than his younger brother, John F. Kennedy, as his father, Senator Joseph P. Kennedy, had groomed him for just such a political career. In the aftermath of this terrible accident, and with little evidence that the other Project Aphrodite flights were having the desired effect, the program was quietly terminated as yet another ill-fated attempt to hasten the end of the war in Europe.

COLD WAR ROBOTIC ADVANCES

After the failure and cancellation of Project Aphrodite, the U.S. military largely shelved its pursuit of remotely controlled weapons. As World War II came to a close in Europe, American personnel raced to capture as many German scientists and engineers as possible, as well as prototypes and finished copies of advanced German weapons systems. Operation Paperclip, the effort to secure as much advanced technology as possible, resulted in major advances in aircraft propulsion and guidance systems as well as rocketry. On the Eastern front, the Soviet Union pursued similar objectives as its forces moved into central Europe. Like the United States, the Soviets hoped to jump-start their military technology programs by pressing German innovators into service.

In particular, the V-1 and V-2 programs showed enormous promise that could not be ignored by any of the Allied powers. The V-1, which had so terrorized England for the last months of the war, led to major advances in the development of air-breathing cruise missiles. The V-2 offered the possibility of placing artificial satellites into orbit, and, if nuclear weapons technology continued to advance, might be used to deliver a city-killing atomic weapon from hundreds of miles away. American missile programs competed for captured German experts, as they possessed not only the practical experience of designing advanced missiles but also the foundational expertise derived from experimentation and failed attempts with previous models. In 1953, Princeton professor of physics John von Neumann predicted that a hydrogen bomb's warhead could be constructed in a sufficiently small size for missile delivery by 1960.[19] Whichever nation first created a missile capable of delivering such a warhead with a modicum of accuracy would derive a major advantage in the burgeoning Cold War. On October 4, 1957, the Soviets demonstrated that they could theoretically send a warhead anywhere on earth when they launched *Sputnik* into orbit. The world's first satellite only weighed 184 pounds, and its primary function was to simply emit a radio signal, demonstrating its presence in orbit for 21 days before its battery ran out—but it had the secondary effect of proving that terrestrial rockets could boost artificial satellites beyond the reach of earth's gravity. It was not a stretch to imagine a warhead being boosted to the very edge of orbital velocity and then being steered back into the atmosphere, to strike elsewhere on the earth with unstoppable speed. There was quite literally no possibility of defending against such an attack—and only the promise of retaliation might serve to curtail the ambitions of a nation that possessed such a capability.

Although the Soviet Union launched Yuri Gagarin into orbit to great fanfare in 1961, the mission was as much about propaganda as it was about scientific achievement. In many ways, robotic platforms offered (and still offer) a far more viable mechanism for space exploration. In 1966, both the Soviet Union and the United States launched robotic platforms to the lunar surface in anticipation of future manned missions. The Soviet *Luna 9* arrived on February 3, 1966, and landed near the Reiner and Marius craters, in the Oceanus Procellarum. It transmitted photographs of its surroundings for three days before its power supply failed. The American *Surveyor I* reached the lunar surface on June 2, 1966, and managed to transmit photos, video, and even some engineering data until January 7, 1967, when it finally fell silent. Building upon *Luna 9*'s success, in 1970, the Soviets launched *Lunokhod 1*, a remotely controlled lunar rover. *Lunokhod 1* roamed over 6 miles across the lunar surface, recording and transmitting data for more than 10 months before finally failing to respond to terrestrial commands. Of course, it arrived more than a year after U.S. astronauts Neil Armstrong and "Buzz" Aldrin, but in many ways it represented a far more practical means of investigating lunar conditions.

Thus far, robotic platforms had proven the only feasible mechanism for visiting the Earth's nearest planetary neighbors. In 1972, after several failures, the Soviet Union succeeded in placing a rover on Venus, named *Venera 8*. After spending 112 days reaching the planet's surface, the lander managed to transmit data for less than an hour before succumbing to the harsh conditions, including a measured temperature of 450 degrees Celsius and an atmospheric pressure 94 times heavier than the Earth's level. Later *Venera* missions (numbers 9–14) also managed to reach the Venusian surface, with each transmitting additional data but none lasting much longer than an hour. In 1973, NASA launched *Mariner 10*, a spacecraft that flew past Venus and Mercury but made no attempt to land on either planet.

There has been a far larger amount of U.S. interest and investment in the examination of Mars, in large part because the planetary environment offers far fewer challenges when compared to Venus. *Viking I* and *II* launched in 1975, and, after spending several months in orbit, both deployed landers to study the Martian surface. Each of the landers sent back an enormous volume of data, and *Viking I* managed to continue its mission until 1982, when a software coding error caused the lander's antenna to retract, cutting all power and communication. NASA did not return to the Martian surface until 1997, when *Pathfinder* landed and set up an automated data station, which incorporated a mobile rover, *Sojourner*. Although they functioned only for a few months, they offered proof of concept of a number of new innovations, enabling later missions to have much greater ambitions. In 2004, NASA launched two spacecraft carrying landers to the Martian surface: *Spirit* and *Opportunity*. Designed to last 90 days, *Spirit* managed to remain mobile for 5 years before finally becoming irretrievably mired in the soft Martian soil. *Opportunity* remains mobile and continues to transmit photos and data more than 13 years after it was scheduled to end its mission. The rovers were joined on the Martian surface in 2012 by *Curiosity*, a semiautonomous vehicle designed to assess the Martian climate and geology in preparation for future missions.

Not surprisingly, robotic spacecraft have presented the only mechanism for examinations into more distant portions of the solar system. In 1972, NASA launched *Pioneer 10*, the first manmade object to leave the solar system. In 1977, NASA followed with *Voyager I* and *Voyager II*, which, due to their higher velocity, will eventually pass *Pioneer 10*. Each initially travelled to examine the gas giant planets. As of 2017, both of the *Voyager* spacecraft have passed beyond the formal range of the solar system. In a remarkable display of engineering prowess, both *Voyager I* and *II* continue to transmit data to receivers on Earth. Each has also continued to respond to commands from NASA, despite the practical difficulties associated with communicating through a computer system more than four decades old. Contact with *Pioneer 10* was lost in 1997, presumably due to power failure, although intermittent signals were received from the vessel until 2003.

In 1989, Rodney Brooks of the Massachusetts Institute of Technology published "Fast, Cheap, and Out of Control: A Robot Invasion of the Solar System." In it, Brooks argued that spacefaring nations should cease their investments in extremely expensive platforms for exploration, a strategy that essentially risked complete system failure on the performance of every component within the platform. To Brooks, this created an exponential risk of catastrophic system failure in an environment where repairs would be effectively impossible. Instead, Brooks believed that nations interested in scientific exploration should create large numbers of infinitely simpler machines, which might then be launched in swarms capable of following independent paths and reporting back the conditions encountered. Even with a failure rate of 90 percent, these swarms would hypothetically collect far more data and do so at a fraction of the resource cost. For some applications of space exploration, Brooks was undoubtedly correct. For example, any attempt to map the location and composition of individual units within the asteroid belt would do well to rely upon the swarming technique. Attempts to examine specific targets, such as planets or their moons, might do better to use more sophisticated machines to allow a wider variety of options when dealing with unexpected stimuli. Thus far, no nation has adopted the Brooks model for space exploration, per se, but it is entirely likely that his approach will prove attractive to terrestrial corporations attempting to discover and capitalize upon raw resources located beyond the Earth.[20]

INDUSTRIAL ROBOTS OF THE COLD WAR

Military robotics and remotely controlled military vehicles became a lower priority for most nations in the early Cold War, although industrial robots began to populate factories around the globe. These robots tended to perform the most basic mechanical functions of assembly-line manufacture, particularly duties that had proven especially dangerous or tedious for the workforce. While they might not offer the same level of intrinsic judgment as human workers, and by definition, did not have the same perceptual capabilities, these machines required only electricity and basic maintenance to perform their functions. They never evinced any desire to join the organized labor movement, never complained about the workplace conditions, and never required a restroom break. In short, to some manufacturers, robotics appeared to be the future of industrial production, a specialized workforce that could keep the factories running at full capacity regardless of outside conditions. In reality, many of the early industrial robots could perform only the most basic of functions and were so expensive that they required years of usage to justify their cost. Further, they proved extremely difficult to repurpose or reprogram, and thus industries that depended upon continual changes to the final product, such as auto manufacturing, found that robots could only replace a small percentage of the workforce. Even the robots that were utilized required a certain degree of human oversight and operation, although this could often be provided by relatively unskilled, and hence much cheaper, laborers.

With all of the concentration upon missile programs, rocketry, and the tangential space race, far fewer resources remained available for investigations into robotics, at least in the military sector. Industrial automation continued apace, as did the development of computers. In 1951, Raymond Goertz designed a mechanical arm incorporating force feedback technology, allowing a distant user to feel when the arm came into contact with an object. Just three years later, George Devol built a reprogrammable robotic arm, allowing industrial applications to modify their behavior to suit new product lines. No longer did an investment in robotics tie a manufacturer to a single model or process, meaning that the automation option had the same flexibility as human workers—but could be reprogrammed in a matter of hours. In 1955, General Motors (GM) installed Planetbot into its Harrison, Pennsylvania, radiator plant. In 1961, the GM assembly line at Ternstedt, New Jersey, began using a Unimate robot to weld die castings onto automobile chassis. In each case, programmable robots took over the most tedious and dangerous jobs within the manufacturing process and performed them more efficiently and effectively than their human counterparts.[21] When Victor Scheinman invented the Stanford Arm in 1969, he revolutionized the development of industrial robots. This entirely electric robot could be programmed to follow arbitrary paths in space along six axes, meaning it could be reprogrammed for a wide variety of industrial applications and it could approach each task from different directions as needed.

COLD WAR MILITARY ROBOTS

When the United States became increasingly embroiled in the Vietnam War, it faced an enemy that had imported a progressively more sophisticated air defense network. Photo reconnaissance missions became extremely dangerous when approaching the most heavily defended targets—and required specially modified fighter aircraft, making each loss exceptionally expensive. In 1962, engineers at Teledyne-Ryan proposed modifying a target drone aircraft, the Firebee, to serve in a reconnaissance function. In particular, this approach provided an effective means to determine the frequency of Soviet- and Chinese-supplied air defense radars. The modified Firebees, rechristened "Fire Flies," were flown over known enemy radar sites in the hope of provoking attacks. When the North Vietnamese radars detected a Fire Fly, they were in turn exposed—and before the Fire Fly could be shot down, it transmitted the radar frequency being used by its attackers. This allowed U.S. engineers to modify jamming devices that could disrupt the radar systems, making U.S. aircraft substantially safer at the cost of a handful of target drones.[22]

Fire Flies were also used to conduct photoreconnaissance missions over heavily defended sites. They rarely survived more than a few missions, as they flew on programmed routes that proved fairly predictable for air defense gunners. Those that survived provided valuable intelligence, but time-sensitive targets still proved problematic, as the Fire Flies' photo negatives had to be recovered, developed, and printed for analysis. To solve this problem, Teledyne engineers further modified drones to carry television cameras and transmitters. These drones, dubbed "Lightning Bugs," could be air launched from under the wings of a DC-130 Hercules and could transmit images back to the controller aircraft, which remained safely beyond engagement range. The Lightning Bugs overflew the Vietnamese prisoner-of-war camp nicknamed the "Hanoi Hilton," verifying the presence of American personnel. Several also flew into Chinese airspace, whether accidentally or intentionally, and were shot down by Chinese air defenses. Upon recovering the crashed aircraft, Chinese personnel were surprised to discover the lack of any pilots, or even a cockpit, making it difficult to argue that the United States had engaged in an act of war against the People's Republic of China.[23]

Specially configured Fire Flies dropped propaganda leaflets upon the Vietnamese civilian population. Occasionally, they dropped chaff as a means to confuse radar sites and open air corridors for manned missions.

The last Fire Flies in the U.S. inventory were actually expended in 2003, when they were equipped with chaff, flares, and electronic "noisemakers." On the first few nights of Operation Iraqi Freedom, they were launched on one-way missions over Iraqi airspace, drawing as much attention as possible from air defense networks. In turn, when Iraqi radars illuminated them, trailing U.S. aircraft attacked and destroyed the radar sites. Iraqi radar operators feared turning on their systems due to the likelihood of destruction, with the result that a few Fire Flies nearly reached the Iraq-Syria border and had to be ordered to self-destruct before violating Syrian airspace.[24]

The success of modified Firebees reinvigorated interest in developing unmanned platforms, at least for intelligence collection missions. In 1986, Israeli Aerospace Industries demonstrated the Pioneer, and after partnering with Aviation Armaments Incorporated, sold the first production copies to the U.S. Department of Defense. Classified the RQ-2, the Pioneer was initially utilized for reconnaissance and artillery spotting, but proved so versatile that it soon performed a wide variety of functions. Only 17 feet wide and 14 feet long, the compact Pioneer could fly to a range of 115 miles at an altitude of 15,000 feet, carrying 100 pounds of cargo. Integrated sensor options included electro-optical and infrared cameras, with the ability to broadcast images directly or through satellite uplinks. It could take off from a runway, be flung into the air via catapult, or use rockets for an assisted takeoff. Landings could be on a conventional runway (or straight section of roadway), or it could fly directly into a shipboard recovery net.

After demonstrating its value in spotting targets for battleships' main batteries, the Pioneer was pressed into service searching for antiship mines in the Persian Gulf. During the Persian Gulf War, Pioneers outfitted with chemical detection and communications interception gear routinely patrolled over Iraqi airspace. Special meteorological Pioneers were used to determine weather conditions along planned flight paths, designating targets for follow-on manned aircraft in the process. On occasion, a Pioneer effectively accepted the surrender of Iraqi troops under battleship bombardment. The Pioneer's sensor operator detected white flags waved by the Iraqis, notified its chain of command, and halted the bombardment in favor of a shore party to formally accept the surrender.

Pioneers continued to serve in the 1990s. During the humanitarian mission to Somali in 1992–1993, Pioneers operated as aerial spotters for gunships flying near Mogadishu. They were also utilized in an effort to locate key leaders of various militias attacking United Nations aid missions. The conditions in Somalia proved almost ideal for the light aircraft, allowing the Pioneers to manage a sortie rate of 93 percent. During the aerial missions over the Balkans, the weather proved far more difficult, as did the relatively sophisticated air defenses used by the Serbian military.

Eighty percent of the Pioneers sent to Kosovo in 1999 were either shot down or lost to operator error. Much like the earlier Fire Flies, the Pioneers in the U.S. inventory flew their final missions in Operation Iraqi Freedom, before being retired in favor of larger and more durable platforms.[25]

The most well-known and successful remotely piloted aircraft to date, the RQ-1/MQ-1 Predator, made its maiden flight in 1994. The aircraft eschewed many of the traditional assumptions about aircraft in the modern era. In particular, it was deliberately designed to fly slow, allowing it to conduct detailed reconnaissance. Because the Predator had no human aboard, and was relatively inexpensive, it was considered an expendable asset. Over the following two decades, Predators became increasingly costly, but not due to the airframe itself. Rather, the primary cost of the Predator came from its incorporated sensor suite, which could include all of the earlier Pioneer's capabilities and far more. Mounted in a specially designed ball at the nose of the aircraft, the sensors can be independently operated from the airframe.

Just one year after the first Predator flight tests, operational units were deployed to Gjadër airbase in Albania, to be flown over the warring Balkan states. Their sensor and transmission capabilities proved invaluable in documenting genocidal atrocities. The early deployment demonstrated that Predators were relatively dependable but experienced significant problems when flown in difficult weather conditions. In particular, the wings tended to ice over, leading to mechanical failures.

LIFESAVING ROBOTS

On September 7, 2001, the world's first telepresence surgery was conducted. Using a robotic manipulation system, Dr. Jacques Marescaux, aided by his French surgical team, was able to operate on a patient in Strasbourg, France, while remaining in New York. Given the transatlantic nature of the operation, it was quickly dubbed "The Lindbergh Procedure" in a nod to the first pilot to fly the same distance as a solo pilot. The robodoc mimicked the surgeon's movements, though it travelled at a much slower speed that nevertheless allowed for incredible precision. This new surgical innovation made it possible for the most talented surgeons in the world to operate on patients anywhere on the planet, opening up enormous possibilities for life-saving operations. It also allowed a specialist surgeon to spend more time focusing upon what he or she had spent years or decades perfecting, rather than requiring hours or days be wasted traveling to remote locations to see a single patient.[26]

Martin Ford argues that this was just the first step in the robotic surgical revolution. Theoretically, a robotic surgeon could perform many routine procedures, without facing the possibility of exhaustion, boredom, or the sudden urge to sneeze at the worst possible moment. Further, if artificially intelligent machines are uploaded with the records of thousands of procedures, they can essentially "perfect" their technique before actually touching a human patient. This same training process for each human surgeon requires years of training at a cost of hundreds of thousands of dollars, whereas the robotic knowledge base can be replicated almost instantly in a limitless number of models. The experience of each surgery performed by a robotic surgeon can also be added to the overall database, meaning that even unusual or unexpected situations might become relatively easily overcome once they have been experienced by a single machine. As a result, each and every robotic surgeon, if continually updated, will remain at the proverbial "cutting edge" of the profession, pun intended.

Telepresence robots are not limited to medical applications, of course, and human-machine teams have enhanced the performance and production of a countless number of occupations. One of the most common such pairings tends to be noticed only in the event of a catastrophe: the incorporation of automation into mass transit. Autopilots have played at least limited roles on commercial aircraft for more than six decades. In 1947, a C-54 military transport made an entire flight, including takeoff and landing, via autopilot, with the human pilot and copilot present only to assume control in the event of an error. This not only demonstrated the capability of the new technology, it also captured the imagination of commercial carriers. Autopilot systems are not required as a part of virtually all commercial aircraft with more than 20 seats and have proven to be much less susceptible to error than humans, although it is not an infallible technology. Autopilots do not fly airlines without human pilots in the cockpit, yet, but they are frequently used for landings conducted under instrument-only conditions. The computer's reaction speed far exceeds that of human pilots, as does its ability to simultaneously receive and process information from dozens of sources. 747 autopilots do not become nervous due to weather conditions or jittery from too much caffeine, and testing has shown that they are far more consistent in routine flight activity. In many ways, the human pilots act as a second opinion and an emergency-management system.

Naturally, human pilots also serve a certain social and psychological function, making passengers feel more secure. However, some of the most recent airline tragedies have begun to influence passengers' feelings about human and computer pilots. On March 24, 2015, Germanwings copilot Andreas Lubitz deliberately locked the rest of the flight crew out of the cockpit of an Airbus A320. He then placed the aircraft in a steep dive that lasted more than 10 minutes. The doomed airplane's flight recorder preserved the increasingly desperate inquiries of Lubitz's fellow aircrew, including a final attempt by Captain Sondheimer to batter down the reinforced cockpit door. Ironically, anti-terrorism security measures prevented him from storming the cockpit and attempting to regain control of the aircraft. At 10:41 a.m., the airplane slammed into the French Alps, killing all 150 passengers and crew. In the aftermath, some experts called for creating an autopilot override system that might sense such a disaster in the making and take steps to prevent it, most likely by seizing control of the aircraft and contacting an authority on the ground for further instructions.[27]

Of course, human-caused mass transit disasters are not limited to aircraft, although airplane disasters tend to grab headlines (and in the process, feed flying phobias.) On September 12, 2008, a metro transit train crashed into a freight train in Los Angeles, killing 25 and injuring 135. The cause of the crash eventually emerged as being due to human error. Specifically, the engineer driving the train, Robert M. Sanchez, could not be bothered to look up from the screen of his mobile device, on which he was sending and receiving text messages in contradiction of company policies. His inattention caused his train to run a red light and strike the freight train head on. The incident demonstrated the utility of turning routine control of vehicles on controlled routes to an artificial intelligence that cannot become bored and that will not ignore company rules regarding paying attention.

THE PUSH FOR UNMANNED ISR

The famous case of Francis Gary Powers illustrates one of the driving factors in the development of unmanned intelligence, surveillance, and reconnaissance (ISR) aircraft. A former U.S. Air Force pilot, Powers accepted a position with the CIA as a pilot for its U-2 spy planes. In that capacity, Powers and others engaged in a series of overflights of the Soviet Union, performing photo reconnaissance missions to map the locations of military installations, industrial complexes, and crucial transportation nodes. In the event of war between the Soviets and the United States, these reconnaissance flights might prove crucial to the success or failure of a strategic bombardment campaign. Naturally, the Soviet Union objected to the flights, which passed directly over its sovereign territory, but it could do little to disrupt the flights, which occurred at altitudes higher than any Soviet interceptor or air defense weapon could reach. For its part, the United States denied responsibility and knowledge of the missions, secure in the assumption that the Soviets could not prove their allegations. The friendly international community did not necessarily believe the American denials but could at least turn a blind eye to the accusations as long as there was no actual proof of malfeasance, and the flights continued with increasing boldness through 1960. The only limiting factor was the range of the U-2, so long as the Soviets did not devise a means of shooting it down.

On May 1, 1960, disaster struck the program. Powers departed an airbase in Pakistan, intent upon a routine overflight. A Soviet MiG-19

attempted to intercept his aircraft but could not reach the necessary altitude. Another aircraft, the newly designed Su-9, reached the necessary
altitude, but because it lacked armament, its pilot could only attempt
to ram the American U-2 and the relative speeds of the aircraft caused
him to widely miss on his only attempt. However, the Soviet air defenses
had recently received a major upgrade in the form of the SA-2 Guideline
surface-to-air missile. Ground personnel fired eight of the new missiles
at Powers, and one hit his airplane. Powers managed to eject safely but
failed to activate the U-2's self-destruct function, and it crashed almost
intact. Powers was quickly captured and subjected to months of harsh
interrogations at the hands of the much-feared KGB.

The official U.S. explanation, that a weather observation aircraft had
strayed deeply off-course and was shot down, was flimsy at best and did
not hold up when the Soviets were able to show the aircraft wreckage,
including the film from its camera, which clearly had not been used to
photograph any weather patterns. Powers proved unable to withstand his
interrogators and eventually made a full public confession of espionage
activities against the Soviet Union, for which he was sentenced to 10 years
imprisonment. Almost two years later, Powers and an American student,
Frederic Pryor, were swapped for Vilyam Fisher, a KGB colonel caught by
the FBI while spying in the United States. The exchange took place at Ber-
lin's Glienicke Bridge, to much fanfare and international press coverage.

The Powers incident was quite embarrassing to the United States—
getting publicly caught in an abject lie often is. Military robots, though,
might be able to perform the same mission with more plausible deniability. When Chinese air defenses shot down modified target drones, they
had no captured pilots to parade before the world. As Paul Scharre notes,

Because of their ability to take more risk, robots could be sent deep behind enemy
lines, not just as scouts but also for intrusive intelligence-gathering and sabotage.
Stealthy uninhabited aircraft can be used for clandestine reconnaissance without
risking a "Gary Powers" incident. While, in the event of a shoot-down or crash, a
highly sophisticated aircraft would not be plausibly deniable, small cheap robots
could be if they were made from commercial off-the-shelf components and without identifying markings.[28]

It is useful to compare the Powers tale with an incident that took place
in the skies over Iran in December 2011. In the later event, an unclaimed
reconnaissance aircraft fell into Iranian hands under disputed circumstances. According to the Iranians, the aircraft was an American reconnaissance vehicle that previously had been flying over Afghanistan,
where it was dubbed the "Beast of Kandahar." They claimed that Iranian
engineers had managed to jam, spoof, or overwhelm the aircraft's command signals and order it to land on Iranian soil, allowing its capture in

functional condition. Initially, the United States denied any responsibility for or knowledge of the aircraft, although it also quietly demanded the return of any malfunctioning equipment. As images of the mystery aircraft emerged, it was eventually identified in the press as an RQ-170 Sentinel, although the United States continued its refusal to publicly confirm or deny responsibility for the aircraft or to explain why it landed in Iranian territory. Experts later opined that while the Iranians could not have actually seized control of the aircraft, they might have jammed its signals, eventually forcing it into an emergency landing procedure when its fuel supply ran low. They also might have spoofed the aircraft's GPS receiver, causing it to mistakenly believe it had left hostile airspace and returned to its home base. Iranian engineers quickly began to disassemble the aircraft and attempt to copy its components. Reports also soon surfaced that Iranian diplomats might have offered to sell the aircraft, as a whole or in pieces, to the Chinese and Russian governments. While the Iranians likely do not have the technical sophistication to replicate the aircraft, the same cannot be said of the Chinese and Russians. Either nation's UAV program would sustain a major boost through the acquisition of the machine.

As long as the overflight of enemy territory is not conducted within the atmosphere, it creates an entirely different set of circumstances. Satellites have flown directly over enemy landmasses since the first artificial satellite launch, *Sputnik*, shocked the world in 1957. Given that the Soviets plotted a course for *Sputnik* that took it directly over American territory, they essentially ceded the argument that such overflights should be prohibited, opening the door for American satellites to begin reconnaissance flights over the Soviet Union. Today, the orbital paths above the earth have become quite crowded, with thousands of working and defunct satellites operating just in low earth orbit. While any nation with the ability to launch a satellite can technically also shoot one down, at least in theory, and, when coupled with nuclear devices, can probably render useless a significant portion of the satellites in orbit, there have been very few attempts to destroy satellites. In 2007, China launched a new antisatellite weapon at a weather satellite in low earth orbit.[29] The result included one destroyed satellite and thousands of new pieces of debris that might offer a danger to other machines in outer space. As if to prove the need for a more effective means of destroying satellites, the United States launched an antisatellite missile in 2008 that knocked a defunct satellite into a decaying orbit that caused it to burn up in the atmosphere, a more elegant solution that left almost no debris while achieving the same effect.

EYES ABOVE THE CONFLICT ZONE

The obvious utility of unmanned platforms for surveillance of hostile and distant zones was not lost on the political leadership of the United

States. Bob Woodward reports that when President George W. Bush first saw the live feed from a Predator, he was immediately struck by its potential applications.

"Why can't we fly more than . . . one Predator at a time?" he [Bush] asked. He had been impressed with the raw intelligence provided by the Predator. It was a useful, low-risk tool—and at a cost of only $1 million apiece, a bargain as far as military hardware went.

"We're going to try to get two simultaneously," Tenet [the CIA head] said.

"We ought to have 50 of these things," Bush said.[30]

Little did Bush know that by the end of his presidency, the U.S. Air Force and the CIA would be flying hundreds of Predators and similar aircraft, and the U.S. Army would possess thousands of smaller tactical ISR aircraft. Writing his autobiography after the end of his second term, Bush recalled the request by the CIA to use armed drones to attack al Qaeda camps and high-value targets. In Bush's words, "George [Tenet] proposed that I grant broader authority for more covert actions, including permission for the CIA to kill or capture al Qaeda operatives without asking for my sign-off each time. I decided to grant the request."[31]

The decision to authorize the CIA to conduct deadly strikes against al Qaeda targets was tactically effective, in that it allowed the rapid use of any new intelligence and surveillance data gathered. But, it had far-reaching implications for not just U.S. policy but the entire world's approach to warfare through remotely controlled machines. Bush's autobiography is somewhat circumspect on this point, but it does note that after authorizing covert assets to engage in lethal operations, the number of enemy combatants killed in airstrikes began to rapidly rise:

I looked for other ways to reach into the tribal areas. The Predator, an unmanned aerial vehicle, was capable of conducting video surveillance and firing laser-guided bombs. I authorized the intelligence community to turn up the pressure on the extremists. Many of the details of our actions remain classified. But soon after I gave the order, the press started reporting more Predator strikes. Al Qaeda's number-four man, Khalid al-Habib, turned up dead. So did al Qaeda leaders responsible for propaganda, recruitment, religious affairs, and planning attacks overseas. One of the last reports I received described al Qaeda as "embattled and eroding" in the border region.[32]

The rise of the killer robots began under the Bush administration, but it reached full fruition when his successor took office.

CHAPTER 4

Automating the Battlefield

Conventionalism takes a firm stand against innovations. But this inflexibility towards any immediate change is often followed in the long run by the most astounding reversals. When it ultimately changes, it often turns upside down and stands on its head.

—B. H. Liddell Hart, 1936

Just as robotics have begun to exert an increasing influence over industrial production and even the delivery of services, so too has there been an enormous move to automate the modern battlefield through the incorporation of increasingly sophisticated robotic platforms. These devices are typically delineated by the primary domain on which they operate, namely land, sea, and air, although there is no fundamental reason that this should be the case. It would be just as simple to differentiate the machines by function, level of operator input and control, or lethality. Yet, the traditional paradigm of separating the military services by primary domain, even though each operates at least to a certain extent in the other domains, is a perfectly serviceable means to break down the very complex subject of modern military robotics on the battlefield.

Although the history of human warfare started on land, spread to the seas, and has only recently entered the air, in many ways, ground robotics have proven the most difficult to develop and field. In part, this is due to the inherently crowded and complex environment in which land operations take place. Wars typically do not take place in conveniently empty locales, and even if they did, there is no guarantee that the terrain would be consistent, predictable, or compatible with robotic locomotion. Just as armies possess and use specialized equipment for various terrain environments, military robots often have a substantial amount of customizability or specialization for different expected locations.

Military robotics might have remained an esoteric, slightly changing field were it not for the wars in Afghanistan and Iraq. The most prominent aircrafts of the War on Terror, Predators and Reapers, were considered somewhat of a novelty in the 1990s. Then-CIA director Robert Gates recalls that the U.S. Air Force had no interest in the platforms, noting "While I was CIA director, in 1992, I tried to get the Air Force to partner with us in developing technologically advanced drones, because of their ability to loiter over a target for many hours, thus providing continuous photographic and intercepted signals intelligence coverage. The Air Force wasn't interested because, as I was told, people join the Air Force to fly airplanes and drones had no pilot. By the time I returned to government in late 2006, the Predator drone had become a household word, especially among our enemies, though the Air Force mind-set had not changed."[1] The change was slow in coming; as Wilson Brissett points out, even in 2007 the Air Force had very few unmanned platforms. "Today [2017], 67 percent of the Air Force ISR inventory is made up of unmanned aircraft. A decade ago, there were only 24 unmanned aircraft in the entire ISR inventory, and they constituted less than 18 percent of the ISR fleet."[2] In 2006, the Air Force possessed 11 RQ-4 Global Hawks—and those were the most prevalent unmanned assets in the USAF fleet.[3] In comparison, at the end of September 2016, the Air Force possessed 33 Global Hawks—but more than 300 Predators and Reapers.[4]

American military forces were almost seduced by the allure of military robots. At last, it might be possible to overcome the fundamental challenges of distance and danger, imposing U.S. will upon enemies without allowing harm to American forces. After all, as Grégoire Chamayou notes, "it is now possible to be both close and distant, according to dimensions that are unequal and that combine a pragmatic co-presence. Physical distance no longer necessarily implies perceptual distance."[5] In theory, military robots allowed the United States to project power anywhere in the world, and to maintain an infinite presence, without all of the normal costs associated with such an activity, making them the panacea desperately desired by political officials who wished to remain the world's sole superpower while avoiding the economic problems and human losses associated with military adventurism.

HOW THE SERVICES VIEW THEMSELVES

Each of the U.S. military services has a distinctly different approach to combat, and in particular in their approach to which members of the service are responsible for the application of lethal violence. In 1989, Carl Builder's *The Masks of War* sought to examine the peculiarities of each service culture, in part by analyzing how each presents itself to the world and in part by assessing how each perceives itself. He distilled the essence of the U.S. Army down to the notion that it considers itself the "servant of the nation." The Navy, in his view, proved almost obsessed with tradition and maintaining its own methodology, including the independent responsibility and activity of ships' captains. The Air Force, to Builder, proved fixated upon cutting-edge technology, almost to the exclusion of all other considerations. Builder also identified the fundamental dividing points within each service, a unique insight that largely holds true three decades later. Within the Army, the key distinction was and is between members of the combat arms (infantry, armor, artillery, attack aviation, and air defense) and the other types of combat support units. Within the combat arms, there is a visible, and occasionally fluid, pecking order, but all of the combat arms maintain a degree of control and preeminence not seen in support organizations. In the Navy, the delineation is between surface warfare, submarine warfare, and aviation. The vast majority of top commanders within the Navy come from carrier aviation, with only a smattering of four-star admirals from the other branches. In the Air Force, the key distinction is between rated officers (those who fly as their primary duty) and nonrated personnel. Among those who fly, there is a pyramidal structure, with pilots inevitably above non-pilots and those who fly the highest-performance aircraft at the very pinnacle of the profession. Ironically, the small numbers of high-performance aircraft, which has in turn kept the number of required pilots for those aircraft low, has created an elite mentality within the group. Yet, the more common types of fighter aircraft, which have seen a much greater utilization level in the 21st century, has led to a significant difference in the number of combat hours flown by F-22 aircraft compared to the less-advanced but more utilized F-16 airframes. Combat hours, which often factor into promotion rates, threaten the preeminence of F-22 pilots and can largely be used to explain the decision to commit the F-22 to the fight against the Islamic State. Otherwise, it would be difficult to justify using a stealthy air-superiority fighter in a ground attack role against an enemy with virtually no air defenses.

Not only does each service have a unique method of presenting itself to the world, each also has a different approach to killing the enemy and who is primarily responsible for lethal action. In the Army, the vast majority of tactical engagements of an opponent are conducted by enlisted personnel. Although an officer is theoretically in command of any combatant unit, in practice, the Army has essentially pushed lethal decision making down to the lowest echelon. In point of fact, if an officer in combat is personally employing a weapon, he or she is probably either in a desperate situation or performing poorly as a combat leader. A private is far more likely to take an enemy life than is any officer. Second lieutenants commanding platoons are taught to lead from the front as a means of inspiring their troops, but a lieutenant colonel commanding a battalion is hardly expected to do the same. If a lieutenant colonel is actively firing a weapon in combat, it is a sure sign that the battle is going very badly for the unit. Taking the concept further, if a lieutenant general is anywhere near the fight, a catastrophe has probably occurred. In fact, only two U.S. lieutenant generals that have died in uniform since World War II both died in non-combat situations. The first, Walton Walker, died as the result of a car accident while inspecting forward positions in the Korean War. The second, Timothy Maude, was killed in the September 11 attacks, when an aircraft struck the Pentagon. The only general officer killed in the War on Terror while serving in a combat zone, Major General Harold Greene, was not killed in combat as well. He was assassinated by a Taliban infiltrator dressed in an Afghan Army uniform.[6]

At the opposite end of the killing responsibility spectrum, the Air Force insists that only officers can be trusted to fly and fight combat aircraft and thus almost all lethal action delivered by the Air Force is done by officers. Most Air Force pilots do not even have the opportunity to command a combat unit until they reach the rank of lieutenant colonel, a promotion that typically occurs after more than 15 to 17 years of service. Thus, the Air Force has a radically different perspective when it comes to responsibility for taking enemy lives. The Navy blends some of the elements of both, in that naval pilots are officers, but surface vessels' weapons crews may be under the control of a non-commissioned officer. In naval combat, there is far less opportunity for decentralized action than what is seen in the other domains and thus the captain of a warship is likely to have much greater control over the utilization of even a single weapon except in the case of aerial attacks.

There is a certain merit to having officers responsible for lethal decision making, particularly when it is carried out in a somewhat sterile fashion (from thousands of miles away). First, officers are at least theoretically better educated, on the whole, than enlisted personnel, and have been specifically trained in the laws of warfare, as well as the process of determining when and how to launch a deadly attack. Second, if anything went wrong in the control center of a remotely piloted aircraft, an officer was more likely to be empowered to take action to correct the problem without seeking higher authority. And third, on average, officers are somewhat older, and hence presumably more mature, than their enlisted counterparts. This perhaps allowed them to more successfully cope with the specific stresses of their roles, especially when noncombatants inadvertently became casualties in the remotely piloted aircraft (RPA) program. Predator pilot Matt Martin described one such accident in vivid detail, when he and his sensor operator carried out a lethal attack upon suspected al Qaeda members in a pickup truck. After stalking the targets for hours, they finally lined up a shot with a Hellfire missile and watched as it neared the target. Just before impact, two young boys entered the kill zone on a bicycle, far too late for the operators to safely divert the missile. In Martin's description, "The truck was a mangled scrap pile of wreckage. The bodies of the two little boys lay bent and broken among the bodies of the insurgents. The responsibility for the shot could be spread among a number of people in the chain—pilot, sensor, JTAC, ground commander. That meant no single one of us could be held to blame. Still, each of us shared in the tragedy."[7]

Martin's description illustrates a larger point regarding the psychological costs of drone warfare as currently practiced. In the words of Mark Bowden, "Drone pilots become familiar with their victims. They see them in the ordinary rhythms of their lives—with their wives and friends, with their children. War by remote control turns out to be intimate and disturbing. Pilots are sometimes shaken."[8] The act of taking another human life is never easy, or at least it never should be, but it appears to be made worse when the attacker must observe his or her target for long periods in advance of the strike and then must loiter in the area, documenting its effects. Military robots are also enabling the documented taking of human lives on an unprecedented scale. Military snipers have often been measured by the number of human lives they have taken—and the deadliest sniper in American history, Chris Kyle, has been credited with more than 160 kills.[9] In comparison, "When Brandon Bryant left the 3rd SOS in April 2011, he was told that over the preceding two years alone, his actions had helped contribute to the deaths of 1,626 enemy combatants."[10]

The Air Force's intransigence stood in stark contrast to the Army's position when it came to new technologies for the War on Terror. This might, in part, be attributed to the dangers of each services' respective responsibilities. When Robert Gates assumed his new position, he saw two fundamental problems, which he considered intertwined. The first was the number of American ground personnel being killed by improvised explosive devices (IEDs). Thin-skinned troop vehicles offered little or no protection against the sophisticated roadside bombs, and troops had begun up-armoring their vehicles using any materials they could grab. Gates managed to untangle a bottleneck in the production of mine-resistant, ambush protected (MRAP) vehicles. The second was the lack of actionable intelligence being gathered from the skies—a low number of ISR assets meant there was little effort to provide an overwatch to convoys moving on the ground. However, whereas the MRAP problem proved to be relatively easy to address, the ISR issue took much greater efforts, largely due to service intransigence. Gates noted, "In the case of the MRAPs, accelerating production and delivery was essentially a matter of empowerment and finding the money. In the case of ISR, I encountered a lack of enthusiasm and urgency in the Air Force, my old service."[11]

The Air Force's insistence that only commissioned officers should fly aircraft, and hence be responsible for attacks upon the enemy, has at times created a frustrating bottleneck for the service and hindered its ability to obtain enough personnel. As former secretary of defense Robert M. Gates noted,

The small number of trained crews available to pilot the drones, particularly in the Air Force, was another significant problem. The Army flew its version of the Predator—called Warrior—using warrant officers and noncommissioned officers.

The Air Force, however, insisted on having flight-qualified aircraft pilots—all officers—fly its drones. The Air Force made clear to its pilots that flying a drone from the ground with a joy stick was not as career-enhancing as flying an airplane in the wild blue yonder. Not surprisingly, young officers weren't exactly beating the door down to fly a drone. When I turned my attention to the ISR problem in mid-2007, the Air Force was providing eight Predator "caps"—each cap consisting of six crews (about eighty people) and three drones, providing twenty-[four] hours of coverage. The Air Force had no plans to increase those numbers; I was determined that would change.[12]

Gates fought a number of bureaucratic struggles with the Air Force, and eventually he removed Secretary of the Air Force Michael W. Wynne and Chief of Staff of the Air Force General Michael Mosely on the same day, in large part because neither one seemed capable of focusing upon the conflict at hand.[13]

The leadership changes appear to have garnered the desired effect. As David Sanger argues,

Today the Air Force has more drone pilots in training than pilots for fighters and bombers combined, a testament to how quickly the culture of the Air Force has changed. A decade ago, drone pilots were regarded as the Air Force's geeks, and flying a Predator from a cubicle was no route to battle decorations. Today the drone pilots hunker down wearing flight suits at Air Force bases from the Nevada desert to the Virginia coast, while their counterparts at the CIA operate their fleet of drones separately. All told, at least sixty military and CIA bases, including the bases from which pilots operate upwards of four hundred Predators and Reapers, make up the vast expanse of the drone program. And a decade ago, almost none of that existed.[14]

In the early days of the War on Terror, there was a significant difference in the RPAs being flown by the CIA and the Air Force: the CIA had embraced the decision to arm the platforms, while the Air Force preferred to use them for their ISR capabilities.[15]

Just as each service's decisions on who should kill has a significant effect upon the way it conducts combat, these same decisions also largely determine which members of a service are put in the most danger, particularly in a conventional war. In the Iraq and Afghanistan wars, most of the Army's casualties came from the enlisted ranks, who are simply more likely to be placed in harm's way. The same has been true for every U.S. conflict involving ground troops—being a private typically means that if you are to engage the enemy, you will be in range of that enemy's counterfire. For the Air Force in the post–World War II era, casualties have been concentrated among the officers, because they are the individuals tasked with flying against the enemy's air defenses and are thus the most likely to be killed or captured. Through World War II, the Navy's casualties were the most evenly spread among the services, because if a warship is sunk, a certain percentage of the entire crew is likely to die in the process. Since World War II, though, most Navy combatants have been aviators flying strike missions—the United States has not had a warship sunk through hostile action in more than seven decades. Thus, like the Air Force, the Navy has seen an increasing casualty rate among its officer class, which has in turn skewed the Navy's approach to combat and lethal decisions.

Why does the question of responsibility for lethal actions matter to the present discussion? It matters primarily because it informs how each service pursues the development and deployment of military robotics. None of the U.S. military services has shown much interest in fundamentally altering the basic tenets of its belief system, the structure of its organization, or the behavior of its combat units. All have sought to incorporate military robots into existing processes and activities, with as minimal a disruption as possible. Certainly, none of the services has demonstrated any interest in making enormous changes in its method of doing business, in large part because the U.S. military has not faced a major battlefield defeat for at least six decades. Instead, the U.S. military has increased its dominant position over potential rivals—witness the rapid destruction of enemy military forces in conventional battles in Iraq, Afghanistan, and the Balkans. In each location, U.S.-led coalitions destroyed every significant concentration of enemy forces that could be located and did so at minimal cost in friendly casualties. In the Balkans, the United States avoided placing its own ground forces in harm's way, while, in Iraq and Afghanistan, at least during the initial invasion campaigns, it performed with such overwhelming firepower that U.S. troops could essentially act with impunity, secure in the knowledge that a massive aerial attack was but a single radio call away.

THE OPENING STAGES OF THE GLOBAL WAR ON TERROR

In the immediate aftermath of the September 11 attacks, President George W. Bush requested authorization from Congress to use military force against al Qaeda, any terror organizations allied with al Qaeda, and any nation that offered al Qaeda sanctuary, supplies, or other forms of assistance. Unsurprisingly, Congress responded with the Authorization to Use Military Force (AUMF) of 2001, which states, in part, that "the President is authorized to use all necessary and appropriate force against those nations, organizations, or persons he determines planned, authorized, committed, or harbored such organizations or persons, in order to prevent any future acts of international terrorism against the United States by such nations, organizations, or persons."[16]

The AUMF passed by a 420–1 margin in the House and a 98–0 margin in the Senate. This law clearly demonstrated that Congress, and by extension, the American people, overwhelmingly supported a campaign against al Qaeda and its affiliates, as well as any nation-state that dared to impose itself between the United States and the object of its retaliation. The single dissenting vote was cast by Congresswoman Barbara Lee (D-CA), who later explained: "It was a blank check to the president to attack anyone involved in the Sept. 11 events—anywhere, in any country, without regard to our nation's long-term foreign policy, economic and national security interests, and without time limit. In granting these overly broad powers, the Congress failed its responsibility to understand the dimensions of its declaration. I could not support such a grant of war-making authority to the president; I believe it would put more innocent lives at risk."[17] Ten members of Congress and two senators did not vote on the measure. The very broad authorization contained within the legislation delivered an enormous latitude to conduct military operations to the presidency, without any specific provision for the end of those powers. So long as the AUMF remains in force, any current president will have many of the powers of a wartime president, even in the absence of a specific enemy nation against which to direct those powers.

Not long after the passage of the AUMF, President George W. Bush deployed forces to invade Afghanistan and topple its Taliban government. For its part, the Taliban refused an American demand to deliver al Qaeda's leader, Osama bin Laden, and his chief lieutenants, who had relocated to Afghanistan from the Sudan in 1996. In reality, the Taliban probably did not have the capacity to comply with such a demand, even if it had been willing to do so in contravention of ancient Pashtun codes requiring the protection of a guest. Although bin Laden was in Afghanistan, he was not under the control of the Taliban, and he was protected by a large force of well-armed al Qaeda fighters. On October 7, 2001, American warplanes, accompanied by remotely piloted aircraft operating in a reconnaissance capacity, commenced Operation Enduring Freedom. After one night of attacks, the Taliban's entire air defense capability essentially ceased to exist, leaving its forces open to continual air attack. U.S. Special Operations Forces, accompanied by similarly elite troops from coalition partners, made contact with elements of the Northern Alliance, a loose collection of tribal warriors opposed to the Taliban. Thanks to the backing of coalition airpower, the main Taliban military forces were quickly routed, enabling an increased number of strikes against al Qaeda targets.[18] As one coalition officer explained, "We could not do what we are doing without the close air support—everywhere I go the civilians and *mujahideen* soldiers are always telling me they are glad the U.S.A. has come. They all speak of their hopes for a better Afghanistan once the Taliban are gone."[19] Much of the air support was made possible through the incorporation of unmanned assets for overwatch, target spotting, and laser designation to guide munitions released from manned platforms.

Within a couple weeks, special forces troops had made contact with the leadership of the Northern Alliance. Although U.S. troops numbered only in the hundreds in the early months of the fight, they brought an extremely sophisticated strategic and tactical planning capability that had an immediate effect upon the beleaguered Northern Alliance forces. They also brought the ability to call in virtually unlimited numbers of airstrikes and to do so in essentially uncontested airspace. The Taliban had virtually no functional aircraft or trained pilots to fly them and very few ground-based air defenses. U.S. fears of man-portable air defense systems, possibly including leftover U.S.-built Stinger missiles used to combat the Soviet invasion of Afghanistan in the 1980s, proved entirely unfounded. Within a few days of aerial attacks, it became evident that the only limitations upon coalition airpower in Afghanistan were the ability to reach the distant area of operations and having enough loiter time to employ a full load of weapons. Airpower offered a decisive advantage that the Taliban could neither offset nor avoid. Northern Alliance forces immediately seized the offensive and engaged in a series of attacks that would have been suicidal without friendly air cover but that proved devastatingly effective when supported from the sky.

The vast distances separating Afghanistan from American airbases and aircraft carriers forced the utilization of massive fleets of aerial refueling aircraft and also increased the general wear upon sensitive aircraft components. Reaching Afghan targets proved plenty difficult; actually, loitering in the region to seek new targets from the air might have been impossible, especially against an enemy alerted to the aerial threat and taking countermeasures of camouflage and concealment. However, the presence of slow, long-loiter unmanned platforms equipped with massive sensor suites greatly improved the situation. Not only could the General Atomics Predator RQ-1 fly over a target area for up to 40 hours, it could also use an integrated laser designator to illuminate a target for manned platforms, which could then swoop in, drop their munitions, and withdraw, leaving the Predators to survey the results of the mission. As Bob Woodward reported, "The agency [CIA] had been operating unmanned aerial vehicles—the so-called Predator drones—on surveillance missions out of Uzbekistan for more than a year to provide real-time video of Afghanistan."[20] The utility of this approach was not lost upon U.S. Air Force operational planners, who immediately called for an increase in the number of unmanned platforms, including a rapid training program to produce enough pilots and sensor operators to support the increase in Predator missions. A major expansion of communications bandwidth also proved necessary—the incredible sensor capabilities required an enormous amount of transmission capacity, particularly if the data were to be beamed via satellite connections.

By early 2002, the Taliban had effectively collapsed as a coherent fighting force. Its leaders had disappeared from public view and had probably fled across the border into Pakistan. Its fighters refused to gather in large groups lest they present themselves as an obvious target for the coalition aircraft constantly circling overhead. The complete lack of any Taliban air defenses allowed coalition aircraft to operate with total impunity—RPAs in ever-increasing numbers began to appear over Taliban and al Qaeda strongpoints, relentlessly seeking targets for their armed, manned compatriots. On a number of occasions, RPAs spotted high-value targets, which managed to escape before they could be engaged by armed aircraft. On one occasion, a Predator team believed they had located Osama bin Laden and a small group fleeing on horseback but could do nothing more than report it to the aerial operations center, which had no armed aircraft within range. Reportedly, the Predator's operators debated crashing their aircraft into the party of suspected terrorists but decided it would be unlikely to actually cause any significant harm and would eliminate any possibility of tracking their quarry's further movements.

EARLY PUSHES FOR ARMED ISR MISSIONS

In and of itself, the difficulty of responding to time-sensitive targets spotted via aerial reconnaissance was not a new problem. Photo reconnaissance aircrafts in every war since World War I have identified targets whose vulnerability to attack was fleeting, and some of those targets could not be attacked before they escaped. To combat this problem, a number of armed reconnaissance solutions had been attempted. For example, during the 1944 campaign against German transportation capacity in World War II, Ninth Air Force's Fighter Command, operating under the direction of Lieutenant General Elwood "Pete" Quesada, detached approximately 20 percent of their fighter aircraft to hunt ground targets of opportunity, in particular locomotives and rolling stock.[21] The A-10 Thunderbolt II, which debuted in 1997, is a purpose-designed antitank aircraft capable of hunting and engaging its own targets. By the time of Operation Enduring Freedom, it had evolved into a highly capable close air support (CAS) platform that could be called in by ground forces engaging enemy units. While it remained on-station, waiting for such calls, it was available to respond to requests from the Predator community for rapid-response attack missions.

Weapons designers had long envisioned a remotely piloted attack aircraft, primarily for two reasons. First, the CAS mission is extremely dangerous relative to other forms of aerial combat. To effectively perform CAS missions, pilots need to fly at low altitudes and relatively slow speeds, making them vulnerable to returning ground fire. Second, the CAS platforms, like other aircraft, possessed limited ranges, whereas remotely piloted aircraft had demonstrated extremely long-loiter capabilities.

On February 16, 2001, engineers at General Atomics test-fired an AGM-114 Hellfire missile at a ground target from under the wing of a Predator aircraft. Some had predicted that the force of the missile's engine ignition might cause the Predator to crash, but the lightweight RPA held up remarkably well. Naturally, the inclusion of two Hellfire missiles, each weighing 70 pounds, added a massive aerodynamic drag upon the Predator airframe, significantly increasing its fuel consumption and lowering its top airspeed.[22] Nevertheless, the integrated sensor suite and guidance systems proved exceedingly capable of guiding a Hellfire to a ground target up to five miles distant. An exciting new dimension of remote warfare suddenly became available to Allied commanders in the field. Armed RPAs roaming over the countryside became increasingly commonplace, blending their traditional role of intelligence, surveillance, and reconnaissance with an immediate ability to engage the enemy. General Atomics, the company responsible for building the Predator, had already commenced work upon a larger model, later dubbed the MQ-9 Reaper. This model, which entered service in 2007, has the capacity to carry a payload of up to 3,000 pounds, which means it can be outfitted with external fuel tanks and a full load of 14 Hellfire missiles. It can also carry much larger ordnance, including laser-guided bombs, and recent models have been tested as platforms for air-to-air missiles.[23]

The destruction of the Taliban regime demonstrated the awe-inspiring power of modern military technology, which can be used to destroy almost any point on the earth, no matter how remote, camouflaged, or fortified. And yet, it also demonstrated a decidedly poor understanding of the region, its geography, and its inhabitants. Coalition leaders seemed surprised that their Afghan allies could not be counted upon to undertake complex military operations with the level of precision expected by Western commanders. When the expected capture of al Qaeda leaders, or recovery of their corpses, did not occur, it became clear that coalition planners had failed to predict that their opponents would flee into the hinterlands and launch an insurgency against occupation forces, rather than staying to fight in a suicidal campaign. Suddenly, what had seemed to be an easy military victory turned hollow—the Taliban neither stayed to face annihilation nor capitulated in the face of a massively superior enemy, and by late 2002, it was clear that the fight for the future of Afghanistan had only begun.

THE PIVOT TO IRAQ

Ardor for the war in Afghanistan quickly cooled in the highest political circles. President Bush and many of his closest advisors soon began to focus heavily upon Iraq, a nation that had languished under economic sanctions and the rigid enforcement of a no-fly zone since 1991. Early in 2003, the administration began making the case for an invasion of Iraq, both to the United Nations and the U.S. population. President Bush argued that Iraq represented a threat to the peace and stability of the Middle East so long as Saddam Hussein ruled the country. Secretary of State Colin Powell made a key case before the UN Security Council that argued Iraq had not abandoned its chemical and biological weapons programs, as promised in 1991, and that it was likely attempting to restart its nuclear program. It was later demonstrated that much of the case presented to the United Nations was based upon faulty human intelligence—the administration believed the outlandish claims of an Iraqi defector over other, conflicting sources of evidence. Tragically, it led the Bush administration to all but abandon the campaign against al Qaeda in Afghanistan in favor of building another coalition against Iraq. The objective for the new alliance was the removal of Hussein, the destruction of any weapons of mass destruction programs in Iraq, and the imposition of a new democratic form of government. It was hoped that a democratic government in Iraq would inspire similar reforms throughout the region and create a permanent partner for the United States within the Arab world.

Remembering the outstanding role airpower had played just a few months prior in Afghanistan, Secretary of Defense Donald Rumsfeld initially demanded a plan to invade Iraq using minimal ground forces (initially two armored brigades) backed by overwhelming aerial force. The planned air campaign, dubbed "shock and awe" by its proponents, was expected to completely paralyze the Iraqi government and military functions, leaving the nation prostrate before the advancing coalition forces.[24] They would, in turn, topple Hussein and his government, quickly organize a new leadership for Iraq, and then evacuate as soon as possible, miraculously leaving behind a stable democracy where the dictator had ruled for the previous 24 years. Army Chief of Staff Eric Shinseki vociferously protested this plan, arguing that while it might successfully reach Baghdad and even possibly capture or kill Hussein, the small ground force would have no hope of imposing order when the Iraqi government collapsed. Shinseki called for a force of at least 400,000 troops, which would still be less than half the size of the coalition army that ejected Iraqi forces from Kuwait in 1991. Rumsfeld eventually agreed to increase the number of U.S. troops headed to Iraq but at nowhere near the numbers Shinseki proposed, and the concession came a cost. Three months after the invasion, Shinseki retired from active duty, a move that many analysts consider the price of clashing with the secretary of defense.

Predators and their fellow remotely piloted aircraft proved very adept at confronting the tactical problems faced by the occupation forces—when U.S. troops started falling victim to a proliferation of buried IEDs, for example, remotely piloted aircraft using infrared sensors soon began to patrol the night skies over major roadways. They quietly scanned the ground below, looking for the telltale sign of an individual or small group placing an IED on or near the road. When it proved impossible to rush ground forces to the area quickly enough to capture them in the act, armed Predators carrying Hellfire missiles were added to the mission. These aircraft could attack the individuals attempting to place coalition forces at risk. Even if only a small percentage of the observers flew armed, they made emplacing IEDs a far riskier activity.

The war against al Qaeda was always in part a war against specific individuals, as the most prominent members of the terror organization remained at the top of the most-wanted list. This designation included a bounty in the millions for information leading to the killing or capture of Osama bin Laden and his chief lieutenants. As the Iraqi insurgency grew, the same model of targeting individuals, particularly enemy leaders, transferred to Baghdad and the surrounding area. U.S. special forces, in particular, spent an enormous amount of time and resources hunting specific high-value targets. These included Saddam Hussein, his sons, and top members of his regime, as well as leaders of al Qaeda's regional franchise, such as Abu Musab al-Zarqawi. As intelligence agencies tried to build an accurate picture of the enemy's human network, military forces relentlessly hunted them, seeking to use each capture for leads to the next quarry. Of course, this approach, which relied heavily upon informants and local intelligence agencies, had significant downsides. As journalist Mark Bowden noted,

As U.S. intelligence analysis improved, the number of targets proliferated. Even some of the program's supporters feared it was growing out of control. The definition of a legitimate target and the methods employed to track such a target were increasingly suspect. Relying on other countries' intelligence agencies for help, the U.S. was sometimes manipulated into striking people who it believed were terrorist leaders, but who many not have been, or implicated in practices that violate American values.[25]

There was simply no mechanism to definitively identify, much less cross-reference, all of the individuals presented to targeting committees.

Individuals identified as members of al Qaeda and its affiliates also landed on a series of lists maintained by the executive branch. These lists included the mildly irritating restrictions of the "no-fly" list, which served the obvious purpose of preventing the individuals on the list from boarding any commercial aircraft. The list proved rife with errors and mistaken identities, but the consequences of being placed on the list remained relatively mild, or at least not a mortal threat. At the opposite end of the spectrum was the so-called kill list, a list that essentially served the same function as a collective 19th century "wanted dead or alive" poster. Although in theory reserved for only the most dangerous actors within terror organizations targeting the United States, the list grew exponentially from its inception. The public has little understanding of the list, how individuals are placed upon it, and perhaps more importantly, if it is possible to get off the list under any circumstances.

One of the most troubling aspects of the kill list is that it is entirely developed and maintained by the executive branch and under a heavy cloak of secrecy. To date, the White House has worked assiduously to prevent any intrusion of judicial or legislative oversight of the list—one of the most consistent policies of the Bush and Obama administrations was to defend the privilege of creating and using the list without external influence. Being placed on the list is akin to a death warrant; it serves as an unpublished notice that the U.S. government has taken a particular interest in an individual and believes that individual has close connections to a terror organization. Because there is no publicly accessible explanation of how it is generated (or even of who is currently on the list), the list represents an extremely troubling aspect of the War on Terror. Strangely, it requires far more judicial oversight (and outright approval) to place a wiretap on a suspected terrorist's phone than it does to target the same person for execution, thanks to the Foreign Surveillance Intelligence Act. Further, the elimination of members of the list is not restricted to members of the military, despite their theoretical monopoly on the use of force on behalf of the nation. Instead, members of intelligence agencies, and even civilian contractors, have intruded upon the lethal activities previously restricted to uniformed military members.

SEND IN THE GROUND FORCES

Despite all of the media attention that has been lavished upon aerial robotics, ground robots probably outnumbered them in the wars in Iraq and Afghanistan. Although exact numbers are almost impossible to pin down, ground robots deployed to those conflicts numbered in the thousands, probably in the tens of thousands. Particularly when IEDs became the most lethal means of insurgent attacks against coalition troops, ground robots became a key implement to counter their deadly effects. The Army has begun using robots for explosives detection and disposal as well. Robot sensors can pick up traces of explosive chemicals and alert human operators to the danger, hopefully allowing human troops to avoid ambushes, mines, and IEDs. Explosive ordnance disposal (EOD) functions can be performed by teleoperating a robot equipped with a camera and manipulator appendages, in a fashion similar to a trained surgeon using a robodoc. While perhaps not every IED can be defused by a robot, the machines can also be used to place explosives for controlled detonation of enemy devices. In the worst-case scenario, enemies might detonate their bombs upon the approach of a robot—something far preferable to losing a human EOD technician. If the EOD robot is equipped with frequency scanners and jammers, it might even protect itself from such a command detonation by preventing the user's signal from reaching the deadly receiver. The two most widely deployed ground robots of the last 15 years are the MARCbot, created by Exponent, and the PackBot, designed and produced by iRobot. Each began in a certain niche, but users found new and creative uses for both as they experimented with wartime conditions.

The MARCbot was built in response to reports that Army troops were investigating caves in Afghanistan by tying a rope around their waists and climbing in, hoping they could be pulled back in the event of an emergency. When General John M. Keene, the vice chief of staff of the Army, heard of this situation, he recommended a rapid acquisitions process to develop a very simple, user-friendly exploration robot. Exponent, Inc., a Palo Alto–based engineering and research firm, responded with the MARCbot in under two months. Essentially, this model is a small wheeled chassis with a single arm holding a digital camera. When it proved successful in cave exploration missions, it went into rapid production and was soon incorporated into units deploying to Iraq who began to face a significant IED threat. As one operator explained the situation in Iraq:

If you suspected that something was an IED, you could: walk up to it and inspect it; drive by it slowly and hope it didn't blow up; shoot at it and see what happened; or use binoculars to view it—all of which were either dangerous or ineffective. The other option was calling the Explosive Ordnance Disposal (EOD) to investigate it with their specialty robots or other methods. However, at that time, EOD response times were often measured in hours, during which the road had to be closed and insurgents could freely set up attacks using mortars and rockets. For warfighters on patrol whose job it was to keep the roads open and safe for travel, none of these options were sufficient or particularly safe.[26]

The MARCbot proved extremely successful and versatile when placed into service. It could be set up for operation in under one minute and operated by any soldier with very little training. It proved extremely dependable, perhaps in part because Exponent deliberately avoided unnecessary features that might complicate the robot and increase its production costs. Over 1,000 MARCbots were sent to Iraq, at a cost of approximately US$19,000 per unit.[27] Because they proved so user friendly, troops tended to make field modifications to their robots. In one notable instance, a patrol incorporated a Claymore antipersonnel mine onto their MARCbots and then used the robot to scout potential hostile locations. If an enemy was detected through the MARCbot's camera, the Claymore was command detonated to eliminate the threat. Although it destroyed the MARCbot in the process, the low cost of replacements made it a palatable situation.[28]

The PackBot, constructed by the iRobot Corporation, is a 42-pound tracked vehicle that is designed to be capable of more complex operations than those performed by MARCbots. It was first placed into the field in 2001, when two PackBots were used to search through debris at the World Trade Center site. By 2002, PackBots had been sent to Afghanistan to assist in scouting and EOD missions. They proved so useful in preventing troops from having to attempt to disarm roadside bombs that U.S. troops refused to return the prototype models first tested. Over the next decade, iRobot supplied more than 2,000 PackBots for service in Afghanistan and Iraq and has continued to develop the capabilities available for the robots to ensure that they will remain a key component of U.S. ground forces for the foreseeable future.[29]

The PackBot is considerably more expensive than the MARCbot, at approximately US$150,000 per unit, but it also has substantially more capabilities. It has eight payload bays attached to the chassis, which allows an operator to swap out modular components to change the capabilities of the basic unit. Not only can the PackBot be used to examine suspected IEDs, certain configurations can be used to actually disarm the weapon, allowing a human EOD technician to remain safely beyond the blast range while working to disable a bomb. The robots proved so effective at neutralizing insurgent traps that insurgents in Iraq began to specifically target PackBots, in the hope that eliminating them would require U.S. troops to be placed at risk.

The PackBot has been tested as a potential ground combat vehicle. In 2011, iRobot engineers experimented with emplacing a shotgun on a PackBot scout model, in part because the shotgun could be loaded with a variety of lethal and nonlethal ammunition.[30] While it has not been used in combat to date, it has also been equipped with an acoustic detection system that essentially allows it to pinpoint the source of incoming gunfire and either transfer that location data or hypothetically return fire.

Both the PackBot and MARCbot have been exceptionally useful to troops in the field, as evidenced by the numbers of each model that have been procured and fielded. Troops have credited both types of robot with saving countless lives, by detecting potential threats before they put friendly troops at risk, defusing IEDs before they can inflict damage, or taking the blast in the place of a human in the event of a detonation. There have been a number of anecdotal examples of troops offering military decorations to their robots, a behavior not previously seen in regards to other military technology. Interestingly, even as the machines are being lauded for their performance, the actual operators of those machines are probably less likely to be decorated for their performance in battle.

The Army has begun to examine lethal ground robots, although it has not committed to fielding any models to date that might take the place of infantry or armored forces. One of the more promising and fearsome such "killer robots" is the Foster-Miller Talon Special Weapons Observation Reconnaissance Detection System (SWORDS). The Talon is a tracked, remotely operated vehicle weighing approximately 200 pounds. It can be used in a variety of missions, including scouting, sensing, and EOD, in a fashion leading some to consider it the PackBot's larger cousin. However, the Talon is a much more rugged machine, designed to withstand at least some combat damage while maintaining functionality. The SWORDS variant can be equipped with a wide swath of potential weaponry, ranging from nonlethal to extremely deadly options. When equipped with an accurate rifle modified with an electronic trigger, the SWORDS has the potential to rival the deadliest of snipers. Its optics and integrated laser range-finder can instantly calculate ballistics tables, adjusting automatically for elevation and windage. It does not suffer from the internal causes of error that plague human shooters, such as a racing pulse, the natural movement of breathing, or the effects of environmental distractions. Sweat will never tickle its nose, and a nearby noise, such as enemy bullets passing nearby, cannot break its concentration. The death of comrades does not elicit an emotional response, nor does the taking of an enemy life. Its sole purpose becomes the detection and elimination of the enemy, in the quickest and most efficient fashion possible. When three armed variants of the SWORDS model were deployed to Iraq, they carried M249 machine guns capable of firing hundreds of rounds per minute. However, they were not autonomous—firing a live round required the release of three different safety mechanisms by the human operator, and there was a single-button kill switch to instantly shut down the robot in case of a malfunction.[31] Even with these safety procedures, the robots were never allowed to fire a round in combat. The reasons for this hesitation have been reported in various fashions, with one prominent article in *Popular Mechanics* essentially claiming that the system was not trustworthy enough to be allowed to fire. The article quoted Army program manager Kevin Fahey: "the gun started moving when it was not intended to move." Fahey also noted, "once you've done something that's really bad, it can take 10 or 20 years to try it again"; precisely what bad event Fahey referred to, however, remains unclear. For its part, QinetiQ, the parent company of Foster-Miller, refuted allegations that there had been any unintended movement of armed variants of the Talon and that earlier malfunctions had been quickly and effectively corrected by changing the hardware of the robot.[32]

Taken to its logical extreme, the Talon presents a terrifying scenario. Imagine a situation in which a ruthless military commander in possession of such robots decides to employ them in a genocidal fashion against an enemy civilian population. Consider that during the Syrian civil war, Syrian president Bashar al Assad showed little hesitation to use any mechanism available to maintain his tenuous grip on power. This included the use of barrel bombs and helicopter-delivered oil drums filled with explosives and shrapnel dropped upon population centers and visible concentrations of citizens. In April 2017, it also emerged that Assad's forces had employed sarin, one of the deadliest nerve agents ever developed, against civilians mixed in with rebel fighters.[33] It is not far-fetched to imagine that he would use deadly robots in the same capacity had they been available to his regime. Perhaps most terrifying, until the outbreak of the Syrian civil war, Assad had been considered one of the most moderate rulers in the region. While not an ally by any means, he was thought to be a source of calming influence who might agree to sign a lasting peace with Israel.

In fact, robots might offer the ultimate method for conducting genocide. Because they can be programmed to kill without concern for ethics or loyalty, they could be turned loose upon a defenseless population on a scale previously unimagined. Many scholars of genocide have pointed out the difficulty of convincing humans to kill on an industrial scale. During the Holocaust, the German military devoted as much as 20 percent of its wartime resources to its extermination campaigns against Jews, Gypsies, homosexuals, political enemies, and other targeted populations. Keeping the Nazi perpetrators of the Holocaust going represented a significant organizational challenge, according to Christopher Browning. In *Ordinary Men*, Browning followed the actions of a single police battalion charged with carrying out the early phases of the genocide and found that many of the men refused to actively participate in mass killings.[34] In comparison, robots programmed to kill civilians would do so with the same lack of compunction as their mechanical counterparts might have while working on an assembly line. They would not question the order to kill, because they have no mechanism or reason to do so.

In 2009, P. W. Singer published *Children at War*, an examination of the rise of underage military forces, particularly in sub-Saharan Africa. Singer argued that children can be induced to engage in horrific acts that adults might resist, in part due to their unformed sense of morality and acceptable behavior. Singer also noted that adult military forces, especially those used by Western nations, have proven particularly resistant to engaging child soldiers on the battlefield, even if they are taking fire from those children.[35] In much the same fashion as the genocidal robots envisioned in aforementioned text, a lethal ground robot will potentially engage any enemy target, regardless of size, age, or gender, and will neither hesitate nor regret its actions if it attacks a young boy or girl wielding a gun.

ROBOTS BELOW THE WAVES

Underwater unmanned vehicles (UUV) have received a lot less attention than their terrestrial and aerial counterparts, perhaps because so few of their operations are carried out in view of the public or the media. Yet, in some ways, the oceans offer a perfect domain for automation, owing to the enormous distances, the relatively predictable environmental conditions, and the enormous number of tasks well suited to robotic platforms. Just as the three primary domains are divided as a means to differentiate operations in each, in the maritime domain, a further discussion is absolutely necessary. In the U.S. Navy, personnel are distinguished as being members of the submarine or surface vessel community. Members of the naval aviation community are often treated as a third group separate from the other two, but for the purposes of this discussion, their contributions to the advancement of military robotics will be relegated to the discussion of the aerial domain.

Military submarines are often referred to as the silent service, referring to the need to control audible emissions if one is to remain undetected below the water's surface. Although it is typical to lump all of the underwater regions of the ocean into the submarine category, in reality, vessels with human crews operate in a very limited region of the underwater domain. During World Wars I and II, thousands of submarine attacks were launched upon unsuspecting surface vessels, sinking millions of tons of vessels and killing thousands of crew members. Countering the submarine threat became a primary consideration for naval strategists, and a limitless number of techniques to counter submarines were attempted. Most met with little success, as detecting a submerged vessel proved extremely difficult. Sonar offered some possibilities in this regard, as did grid-pattern attacks with depth charges. Most submarines could not withstand even a near miss from a depth charge, as the explosion could rupture the submerged vessel's hull and allow a fatal inrush of seawater. Interestingly, during World War I, submarine warfare was conducted at depths of only up to 150 feet of water, the deepest operating depths open to submarines of that period. During World War II, most submarines could dive to only 400 feet, despite two decades of effort to improve designs to allow much greater operating depths.[36]

The average depth of the Atlantic Ocean is 11,000 feet; even reaching that level would open up tremendous possibilities, both for development and for military conflict. Given that the deepest measured point in the world's oceans, Challenger Deep in the Marianas Trench, reaches at least 36,000 feet below sea level, it is clear that there is a much larger area potentially available for underwater operations if the technological problems associated with enormous pressures can be solved. The very first scientific expedition to reach the seafloor in the Challenger Deep occurred in 1960, when Jacques Piccard and Don Walsh rode in the bathyscaphe *Trieste*. They observed sole and flounder at the very bottom of the sea during their 20 minutes at maximum depth. A later expedition (the Japanese remotely operated vessel *Kaiko*) spotted other organisms at the same depth. In 2012, Hollywood superstar director James Cameron piloted *Deepsea Challenger* to the bottom of Challenger Deep, making it only the second manned expedition, and managed to stay at the bottom of the trench for more than three hours. Reaching such depths is certainly not impossible, as a diverse array of species are known or suspected to thrive at such depths; perhaps biomimetics will offer solutions to these vexing problems.[37]

Returning to the matter at hand, that of modern underwater robotics, it should be obvious that unmanned vehicles offer enormous possibilities to reach depths unthinkable for inhabited vehicles. These devices can be purpose built and obviously do not have the complicating factor of having to maintain life support systems or even be constructed large enough for human occupants. Just as surface vessels in the 20th century struggled to detect and counter undersea belligerents attacking from below, so too will modern submarines face a similar threat from below via unmanned submersibles. These devices can be built smaller and much cheaper than military submarines, while still presenting a deadly menace to everything above them. While two manned expeditions reached the bottom of Challenger Deep over a period of more than five decades, two unmanned expeditions have reached the same depths in the past decade, and dozens of research teams are now specializing in remote investigations of the deepest points in the oceans.

The biggest hurdle in the development of truly deep submersibles is the command and control problem. Early iterations of unmanned aerial vehicles had a range limited by the line-of-sight transmission of control signals, although this limitation was soon overcome by satellite links. The command and control of 20th-century submarines presented its own problem, as radio waves do not travel well through water. This forced submarine captains to surface their vessels if they wished to check in with higher authorities. The same problem applies to modern UUVs, particularly if they are to be remotely operated. Instead, it will almost certainly require that these machines operate in an autonomous fashion, lest they give themselves away by surfacing to communicate with their controllers. Granting autonomy to UUVs, if coupled with weaponry, could lead to an unexpected attack, which could in turn trigger an unwanted but unavoidable escalation cycle leading to a general naval war.

UUVs might serve a purpose very similar to that performed by land-based EOD robots, namely, the detection and neutralization of underwater mines. As one theorist noted,

One promising avenue for further exploration is the use of robotics for countermine amphibious operations. Deployed from large-diameter uninhabited underwater vehicles, submarines or surface boats, amphibious robots could find and clear beach obstacles and mines prior to the arrival of amphibious assault troops. Once ashore, robots could establish a perimeter and act as scouts and sentries for the amphibious assault itself.[38]

The notion of using robots to assist in amphibious assaults is not far-fetched—the Talon SWORDS, for example, is already capable of operating in up to 100 feet of water, well beyond the requirements for most coastal approaches.

GETTING BACK TO THE OLD WAR TRADITIONS

In some ways, shifting to robots facilitates a return to the old American habit of a small standing army, massively expanding when the fight is under way. For example, at the commencement of the Civil War, the entire U.S. Army stood at barely 20,000 officers and enlisted troops, with most of them stationed throughout the western frontier. President Lincoln issued a call for state service volunteers that eventually brought more than two million troops into Union service. The Confederacy followed a similar model to raise more than 800,000 troops for service in the war. During World War I, when the United States declared war upon the Central Powers in 1917, it possessed a regular force of only 200,000, despite the obvious warning provided by the first three years of war in Europe. In only 18 months, however, that number swelled to more than four million. World War II erupted in 1939, yet when Japan attacked Pearl Harbor at the end of 1941, the U.S. Army stood at only 1.4 million, even though President Roosevelt considered American entry into the war almost inevitable.[39] By the end of the conflict, more than 16 million Americans had been inducted into military service. In the Cold War era, the United States maintained a much larger standing military force than it had ever previously possessed, relative to the size of the population. This may have had the paradoxical effect of encouraging American interventionism in conflicts around the world—the United States got involved in Korea and Vietnam in part because it could do so without requiring a declaration of war to stand up sufficient military force.

If in the modern environment, the United States does not have the luxury of building a force after getting into a conflict, military robotics might create a standing force that can theoretically always be available. American military forces now stand at their lowest personnel levels since World War II, and given the rise in the size of the U.S. population since that war, the percentage of Americans in uniform has declined at a precipitous rate. To counteract the numerical disparity between U.S. forces and their most likely opponents, American strategists have increasingly turned to advanced technology. Better equipment served as a force-multiplier, making each human in the military capable of more destruction and a wider variety of activities. However, this has also required increasing specialization of troops and a longer period of training before they are capable of functioning on the modern battlefield. It has also made military equipment significantly more expensive, in turn greatly reducing the number of copies procured of the most advanced equipment. For example, the B-2 bomber, first fielded in 1997, cost approximately US$1.157 billion per airframe, with total procurement costs averaging US$2.1 billion per copy built. This price tag caused Congress to limit the U.S. Air Force to purchasing only 21 airframes, rather than the 132 initially envisioned.[40] While it is capable of extremely impressive performance, every B-2 lost represents an irreplaceable cost to the force, and losing only a few in a short period would cripple the long-range strike force. Given that nearly 53,000 American aircraft were lost during operations in World War II, it is clear that the new dependence upon high technology has forced major adjustments in how the United States fights its wars.[41] Ground equipment has also become far more advanced—and expensive—than the mass-produced vehicles of past conflicts. This means that the United States likely cannot fight a war of attrition in the future, as replacing materiel losses would be too slow and costly in an era of rapid conflict.

For the past several decades, the U.S. military has pre-positioned equipment at a number of locations around the globe, ready to be unpacked and put into immediate service in the event of a crisis. This should not be confused with "mothballing," a practice by which naval vessels are preserved in a state of semi-readiness, which would require a few months before they could be flung into conflict. Pre-positioned equipment is usually available for use within a matter of days, not only allowing for a rapid reaction force to operate virtually anywhere on the planet but also ensuring that the cost and time associated with transporting armored vehicles and other platforms to a war zone will be greatly reduced. Military robots might take the concept of pre-positioned equipment to a more extreme situation, in part because they might be capable of unpacking and transporting themselves to the site of a conflict. Just as Amazon relies upon robots within its enormous warehouses to select and pack goods for shipment to online customers, a military robotics warehouse might depend upon a few active robots to select and prepare a specific robotic strike package to fulfill a military mission.

Returning to the theme of incorporating battlefield robots, each of the military services has somewhat struggled to adapt to the latest technological developments. Replacing an infantry soldier with decades of experience in making snap judgments is no easy task, particularly when it comes to using force. Some of the most promising Army robots are not designed to kill, so much as to enable the human killers they accompany to do the killing and at times to protect them from harm. For example, the BigDog robot, initially built by Boston Dynamics, which was purchased by Google in 2013, is for all intents and purposes a robotic mule.[42] It can carry more than 300 pounds of gear and move at the speed of dismounted infantry troops, including over very rough terrain.[43] Given that 21st-century combat troops often move while encumbered by up to 100 pounds of gear, such a load-carrying robot might offer increased agility to fielded forces. There is even the possibility of using such a robot for emergency casualty evacuation—once stabilized, a wounded soldier could be strapped to the robot and sent to a predetermined collection point. If the robot can be correctly and clearly marked, it might even claim protections under the laws of war that prohibit firing upon ambulances and medical personnel.

There have been a number of reports of Army personnel becoming particularly attached to the robots assigned to their units. In part, this is no doubt due to the fact that so many lives have been saved by the incorporation of EOD robots. However, there also seems to be a decided tendency to anthropomorphize military robots. As Wilson Wong writes, "We have seen other cases, of course naming them, giving them ranks, taking risks for the robots that they really should not . . . There was one incident where a robot was stuck and a soldier in Iraq ran out 50 meters under heavy machinegun fire to rescue his robot. The whole point of us using robots in war is to limit risks, and yet here he was taking far greater risk to rescue it."[44] If troops become so attached to their robots that they will risk their own lives to protect them, the value of incorporating robots into the force rapidly declines.

Of course, the Army tries not to walk into enemy ambushes, and in this regard, scouting robots have proven extremely helpful. The most common aircraft in the U.S. Army inventory is the AeroVironment RQ-11 Raven, a small hand-launched tactical surveillance robot. Ravens can be controlled by a single operator using a laptop computer. They can fly for up to 90 minutes, at a height of 500 feet, all the while broadcasting images back to the operator from the small camera mounted in the nose of the tiny airplane. This allows a simple "over-the-hill" look ahead of troop movements.[45] Ravens proved so popular in Iraq and Afghanistan that the Army has purchased at least 10,000 of the small vehicles since its introduction in 1999. Each Raven relies upon a small rechargeable battery to power its flight, which minimizes the time between flights if backup batteries remain available.[46] Landing the Raven is extremely user friendly, as it essentially glides back to earth for a belly-landing. Each Raven is expected to make 200 flights before gradual wear and tear renders it unserviceable, meaning each Raven flight costs approximately US$400.

The U.S. Army has a long-established mistrust of the U.S. Air Force regarding the direct support of ground forces in combat. When the Air Force separated from the Army in 1947, aerial commanders fixated upon the concept of strategic attack and the delivery of nuclear weapons. Many air commanders thought ground forces would play a greatly reduced role in future wars and thus any focus upon CAS or scouting for ground troops would be a waste of resources. In 1966, the Army essentially agreed to forego fixed-wing aircraft, although it reserved the right to build and field rotary-wing aircraft (helicopters).[47] The advent of unmanned platforms has caused the Army to rethink its position on the issue and to procure variants of ISR aircraft flown by the Air Force and intelligence services. Renaming a Predator into a Gray Eagle does not make it a fundamentally different aircraft, especially given that it is produced by the same corporation, in the same factory, using the same components. Essentially, Army leaders do not think the Air Force will choose to support ground forces if budget constraints force a reduction in Air Force platforms. The 21st-century fight over the A-10 Thunderbolt II lends weight to the Army's argument. This purpose-designed CAS aircraft has been used for ground support since its introduction into the Air Force inventory in 1977. In 2013, Air Force leaders announced that the airframe might be retired, largely as a cost-saving measure. The Army initially suggested interest in transferring the A-10 inventory to its control, a move that would obviate the 1966 agreement on fixed-wing aircraft but that would also theoretically involve the transfer of Air Force A-10 pilots to the Army.[48] When the Air Force refused and insisted the aircraft could be decommissioned, A-10 proponents turned to the media and political supporters. In 2016, the Air Force announced the A-10 would be retained indefinitely, despite efforts to shift operations to more multirole aircraft that it insists will be able to do the job just as well (although at a much higher cost per hour flown).[49]

THE RISE OF MERCENARY FORCES

Mercenary forces have been a part of human conflict for thousands of years. In some eras or geographic locations, they have served as the predominant military units. Naturally, they provide the ability to rapidly increase available military power to any employer who can afford their services. The rise of nationalism and the corresponding growth of mass armies, especially through conscription, led to a temporary reduction in the utility of mercenary units. In part, this was due to economic reasons—why pay the expensive rates charged by military professionals when one can create an enormous mass of poorly trained amateurs instead? The advent of industrial production of firearms and ammunition made their human wielders far less specialized, essentially reducing them into another military commodity required to keep the machine advancing. In part, mercenary units declined because in the largest conflicts, virtually every able-bodied male was pressed into service, leaving few behind to potentially become members of a mercenary unit. In the 21st century, mercenary forces seem to be on the rise once more, particularly in nations without a strong tradition of military subordination to civil authorities.

Mercenaries fell out of fashion as the concept of fighting on behalf of one's country as a condition of citizenship became the norm. Throughout the 19th century, they gradually became less common, and, as international customs became codified into the first laws of armed conflict, mercenaries began to be classified as a different form of combatant. This differentiation soon developed into an open prohibition upon the use of mercenary forces. Such groups were perceived as little more than armed ruffians seeking any opportunity to loot and pillage.

By the time World War I erupted, warfare had long since evolved into a titanic struggle between nation-states. Any pretense at victory through maneuver quickly dissolved in the clash of massive conscript armies battering away at one another in a grisly struggle of attrition. The killing and maiming of an entire generation of European youth continued on an industrial scale while political and military leaders experimented with innumerable schemes to drive more troops into the Western front's meat grinder. The result was simply more bloodshed and little if any tangible effect on the positions of the lines. Mercenaries played no significant part in the conflict, although colonial conscripts certainly contributed to their imperial overlords' chances of victory. Any professional fighter not bound to a warring state who possessed a degree of sense or an instinct for self-preservation would not dare risk his person for mere money in that war. If anything, World War II magnified the concept that everyone had a role to play in a conflict, and none of those roles would be mercenary in nature.

Yet, mercenaries did not completely disappear from conflicts in the first half of the 20th century. Rather, they moved to areas where they might enjoy a tactical or technological edge over opponents. Wars conducted outside of the heavily industrialized nations still might be fought by professional freebooters rather than uniformed troops. Also, as large corporations moved into the resource-rich but politically unstable portions of the globe that had been relegated to colonial status, they often brought private mercenary armies to enforce company rules. Little protest erupted so long as they did not engage in fights with the duly-recognized troops of the distant sovereign country, and almost no thought or sympathy was wasted upon abuses of local native populations.

During the Cold War, mercenary units continued to flourish, particularly in third-world nations. As the decolonization movement swept through Africa, Asia, and the Middle East, it was often driven forward in part by mercenary units fighting for money rather than any political ideology.[50] Even after the demise of the great European empires, the geographic boundaries of newly independent states remained somewhat fluid and open to exploitation by decisive rulers. Mercenaries could often be utilized to tip the balance of ongoing conflicts, whether they involved interstate feuds or internal civil wars. Especially when a nation discovered an exploitable natural resource, such as a vein of precious metal or a deposit of gemstones, mercenary warriors could be almost guaranteed to arrive shortly afterward to share in the newfound riches. The continual rivalry between NATO and Warsaw Pact nations might have triggered a buildup of conventional forces in Europe, but it did little to tamp down mercenary activity. Also, veterans of military forces on both sides of the Cold War soon found that their training, familiarity with advanced weapons, and potential access to black markets for arms and ammunition made them a very desirable target for mercenary recruiters. This proved especially true for any veterans of special forces units, whose training prepared them for the harshest environments and the most unconventional tactical situations. Joining such a unit offered many of the attractive aspects of military service without the disciplinary measures common to most nations' armed forces. It also promised a much more lucrative return than the pay rate provided by any national military and the possibility of a much more exciting lifestyle than the one supplied by a peacetime conventional force.

When the slow collapse of the Soviet Union began in 1991, Soviet veterans flooded into mercenary units, armed primarily with black market weapons supplied by corrupt former Soviet officials. P. W. Singer noted that by the mid-1990s there were dozens of major mercenary organizations operating in Africa.[51] The largest source of mercenary troops on the continent, the Republic of South Africa, formally banned membership in mercenary groups in 2006, but the organizations merely shifted their home bases and continued to offer military solutions to political problems, so long as an employer could afford their ever-mounting fees. In particular, South African mercenary units have proven particularly effective at taking direct action against Boko Haram, a terrorist organization primarily located in Nigeria.[52]

A cloak of respectability has been drawn around mercenary units in the 21st century. The term "military contractor" has largely displaced "mercenary," and the modern mercenary organizations tend to shy away from traditional combat roles in favor of more specialized functions. In particular, personal protective details of key political figures in combat zones are often supplied by such contractors, many of them made up of former special forces members. These individuals can command the highest payouts thanks to both their extensive training and the unique status their presence conveys. In the 2003 invasion of Iraq, military contractors were embedded within coalition troops, playing a wide variety of roles. In theory, they were not used for conventional combat and were held to strict rules of engagement regarding the use of force. In practice, they were often reported to be trigger-happy cowboys who felt no need to follow the established laws of war. Because the Iraqi government struggled to exert even a semblance of control over its sovereign territory, mercenary units seemed to operate under the assumption that they had no need to respect Iraqi laws. On numerous occasions, military contractors were accused of using excessive force or even murdering civilians, but they faced almost no consequences for their outrageous behavior. In one terrible incident, Iraqi insurgents killed four contractors working for a private military firm Blackwater USA and dragged their bodies through the streets of Baghdad before hanging them from a highway overpass.[53] On May 8, 2004, the body of contractor Nicholas Berg was found on an overpass in Baghdad. He had been publicly executed by Abu Mussab al Zarqawi, who decapitated the screaming Berg and broadcast the act on the Internet.[54] The act was deliberately conducted to shock and horrify audiences throughout the world and to demonstrate the resolve of al Qaeda adherents fighting in Iraq.

THE TROUBLE WITH TREASON

In the 21st century, at least for the United States and most Western nations, warfare is almost inextricably linked with terrorism. In particular, most Americans assume that the majority of conflicts are conducted in the Middle East, or at least in regions with a majority Muslim population, and that those fights pit a technologically savvy U.S.-led coalition against ignorant savages fighting in pursuit of an evil cause. Of course, there are a host of problems with these assumptions, but popular misconceptions tend to be exceedingly difficult to combat, particularly when they cross ethnic, geographic, religious, and cultural boundaries. It is far easier to demonize the far enemy than to study him, and, to a certain extent, political demagogues utilize that process to harness the energy of their constituents and use it to fuel the conflict. This behavior is nothing new—in virtually every war, there is a substantial degree of misinformation between opponents, particularly regarding motivations and behavior, even when the belligerents have a shared cultural heritage. Civil wars typically erupt in locales with such a shared heritage and yet often devolve into the most barbarous of wars, with atrocities—whether real or imagined—provoking retaliations and escalations, creating a spiral of increasingly destructive behavior.

Americans are usually shocked to discover that the population might harbor citizens who identify with the opposing side in a war. The effort to root out such fifth columnists, often under the charge (or suspicion) of treasonous activities, tends to absorb an enormous amount of wartime resources. This is a natural process—a well-placed traitor really can do enormous harm to a nation's war effort, and leaving such traitors undetected can only enhance the amount of harm they might accomplish. During the American Revolutionary War, Major General Benedict Arnold felt slighted by the Continental Congress when other officers advanced in rank ahead of him. Gradually, this grievance expanded, until he eventually made the fateful decision to sell to the British the fortifications under his control at West Point, New York. Although the attempt failed, Arnold's name is still synonymous with treason in the United States. Even the monument to his greatest battlefield victory in the war, at Saratoga, does not bear his name. Instead, it shows a relief carving of a man's leg and the following inscription: In memory of the "most brilliant soldier" of the Continental Army who was desperately wounded on this spot. Arnold's leg, lost at Saratoga, is the only portion of him that never turned traitor and thus was the only portion celebrated in the macabre monument.[55]

Treason carries a special weight in American culture, one that has existed for centuries. It is the only crime that merits special discussion in the U.S. Constitution, with Article Three declaring: "Treason against the United States, shall consist only in levying War against them, or in adhering to their Enemies, giving them Aid and Comfort. No Person shall be convicted of Treason unless on the Testimony of two Witnesses to the same overt Act, or on Confession in open Court. The Congress shall have Power to declare the Punishment of Treason, but no Attainder of Treason shall work Corruption of Blood, or Forfeiture except during the Life of the Person attainted."[56] Given that the Constitution trumps all other sources of law in the United States, and that no amendment had addressed any aspect of treason, it seems obvious that any conviction for committing treason (and subsequent punishment) can come only after a criminal trial. Further, if and when an individual is convicted of treason, the punishment, specified by Congress, should fall upon that individual alone and not upon that person's entire family.

Even in this highest of crimes, U.S. citizens possess certain inalienable rights of protection against the activities of the government. In particular, the Bill of Rights contains several significant passages regarding the rights of Americans accused of crimes, up to and including treason. For example, the Fifth Amendment states:

No person shall be held to answer for a capital, or otherwise infamous crime, unless on a presentment or indictment of a grand jury, except in cases arising in the land or naval forces, or in the militia, when in actual service in time of war or public danger; nor shall any person be subject for the same offense to be twice put in jeopardy of life or limb; nor shall be compelled in any criminal case to be a witness against himself, nor be deprived of life, liberty, or property, without due process of law; nor shall private property be taken for public use, without just compensation.[57]

Interestingly, the military exception is less well defined. It might exist due to betrayals like Arnold's, still fresh in the minds of many delegates to the Constitutional Convention (having happened less than a decade before the Bill of Rights was drafted). It might be present due to the inherent pace of military operations—a traitor in the ranks might do irreparable harm if he or she is permitted a long legal process. Or, it might be simply a recognition that the military has its own justice system that might diverge at times from civilian procedures. Given the size of the U.S. military in 1791, when the Bill of Rights was adopted, the presence of even a single traitor might have enormous ramifications. Further, the amendment is extremely clear that an individual who stands accused of a crime cannot be punished until that individual has been convicted in a court of law.[58]

Criminal defendants charged with treason are also protected by the Sixth Amendment: "In all criminal prosecutions, the accused shall enjoy the right to a speedy and public trial, by an impartial jury of the state and district wherein the crime shall have been committed, which district shall have been previously ascertained by law, and to be informed of the nature and cause of the accusation; to be confronted with the witnesses against him; to have compulsory process for obtaining witnesses in his favor, and to have the assistance of counsel for his defense."[59] While some exceptions have been granted regarding the definition of a "public trial," and federal crimes such as treason can be prosecuted in any state or district, citizens accused of treason still must have an opportunity to present their case at trial after hearing the case against them—this vital protection requires the government to lay out its entire case prior to offering any defense.

These amendments, as well as a host of federal statutes and procedures, safeguard American citizens from their own government, which has certainly been guilty of overreach on many occasions. This is not to say that traitors have never been eliminated outside of the traditional courts, particularly in military settings. In 1847, General Winfield Scott ordered the execution of 50 captured members of the San Patricio Battalion, a unit fighting for the Mexican government that was composed almost entirely of American deserters, most of them recent immigrants from Ireland. Interestingly, those who had deserted the U.S. Army prior to the formal declaration of war were punished by branding, meaning a red-hot iron poker was pressed into their cheek or forehead to create a burn scar in the shape of a "D," forever marking the miscreant for his crime. The series of hangings of the members of the San Patricio Battalion, conducted from September 10 to 13, 1847, comprised the largest mass execution in the history of the federal government, yet it remains a largely forgotten episode for all but specialists.[60] American soldiers have been executed for other crimes, including desertion, but openly fighting for the opposition tends to provoke the strongest sentiments of condemnation.

On November 25, 2001, U.S. expatriate John Walker Lindh was captured among a band of Taliban fighters. He was detained at Qala-i-Jangi fortress, where Taliban detainees soon staged an uprising against their Northern Alliance captors. In the ensuing fighting, Lindh was shot in the leg and eventually recaptured. Upon interrogation, he admitted that he was an American citizen, a revelation that triggered an FBI interrogation and the reading of Lindh's Miranda rights. The fact that an American citizen, particularly one raised in a Catholic home, might convert to Sunni Islam and join the Taliban and associated terror organizations sent shock waves through the United States. Many called for his immediate execution as a traitor, regardless of any Constitutional prohibitions of such an action. In February 2002, a grand jury returned an indictment against Lindh on 10 charges, including conspiracy to murder U.S. citizens, although a charge of treason was not brought against him. Lindh initially pled not guilty to all charges but then reversed himself and pled guilty to two charges in exchange for a prison sentence of 20 years (rather than the three life sentences his initial charges might carry). Part of the plea deal included that Lindh not speak about any of his experiences and that he drop all allegations of mistreatment at the hands of the U.S. government officials.[61]

Other American citizens caught up in treasonous activities during the War on Terror have not fared so well, especially those with a background radically different from that of most U.S. citizens. In particular, the case of Anwar al-Awlaki is an illustrative example of one of the most complicated aspects of the War on Terror, which at the same time demonstrates many of the key aspects of robotic warfare in the 21st century. As such, his experience bears significant scrutiny, as it might represent a harbinger of the future. Also, his story diverges so significantly from the John Walker Lindh case as to suggest that there might be different standards for different citizens, even if they are accused of the same activities.

Anwar al-Awlaki was born on April 21, 1971, in Las Cruces, New Mexico, and as such had an inalienable right to claim U.S. citizenship. He was raised in Yemen but returned to the United States to pursue higher education. In 1994, he completed a bachelor's degree of civil engineering from Colorado State University, although he did not ultimately pursue an engineering career. Instead, al-Awlaki chose to follow a religious path, serving as the imam of a mosque in San Diego from 1996 until 2000. At the time of the September 11, 2001, attacks, he was an imam at a Falls Church, Virginia, mosque, where he developed a reputation for preaching in both English and Arabic, bringing a number of non-Arabic speakers into the fold. Al-Awlaki was the first imam to conduct a prayer service for the Congressional Muslim Staffer Association and was also invited to conduct a prayer breakfast at the Pentagon shortly after the September 11 attacks. In a classic example of different branches of the government failing to communicate, at the same time he conducted these prayer sessions, al-Awlaki was being investigated by the FBI for potential connections to the hijackers. Several of the 19 men involved had attended prayers at the Falls Church mosque, although there is little evidence that al-Awlaki had any direct connection to the attackers, who maintained a strict operational security protocol. There has been no evidence released publicly to demonstrate that al-Awlaki had any direct knowledge of the plan to hijack four civilian airlines and deliberately crash them into economic and military targets.

Fearing that he might face prosecution from the federal government, al-Awlaki chose to undertake a lecture tour of the United Kingdom from 2002 until 2004; at the end of it, he accepted a position at Iman University in San'a', Yemen. His position in Yemen, far from the reach of the U.S. government, afforded him the opportunity to expand his preaching and more openly attempt to radicalize other Sunni Muslims. Al-Awlaki proved extremely effective at inciting young Muslim men and inducing them to volunteer for jihadist activities. He was especially well received among American Muslim audiences, probably in part due to his intimate understanding of U.S. culture and the challenges it creates for young Muslim men. Recordings of al-Awlaki's sermons and personal communications with him were discovered among the personal effects of al Qaeda planner Ramzi bin al-Shibh, leader of the 9/11 attacks Mohamed Atta, and attempted Times Square bomber Faisal Shahzad, among others. He was also shown to have directly communicated with Fort Hood gunman Nidal Hasan, although there is no proof that he directly ordered any specific attacks. Al-Awlaki also had a significant social media presence, particularly in the form of YouTube videos, which proved extremely popular with his target audience. Copies of his radicalizing sermons circulated among at-risk youth and his ability to shift seamlessly from Arabic to English resonated with American youth.

Al-Awlaki's radical incitement, coupled with his open connection to an al Qaeda franchise in Yemen (al Qaeda in the Arabian Peninsula, or AQAP), led U.S. Treasury Department officials to place al-Awlaki on the Specially Designated Global Terrorists list on July 16, 2010.[62] He had the unfortunate distinction of being the first U.S. citizen to be placed upon the list, though he was certainly not the first one to be tracked and targeted by American forces.[63] Four days later, the United Nations formally added al-Awlaki to the United Nations Security Council Resolution (UNSCR) 1267 list. Writing well after the fact, then-CIA director Leon Panetta defended the decision to target an American citizen by stating, "[al-Awlaki] actively and repeatedly took action to kill Americans and instill fear. He did not just exercise his rights of free speech, but rather worked directly to plant bombs on planes and in cars, specifically intending those to detonate on or above American soil. He devoted his adult life to murdering his fellow citizens, and he was continuing that work at the time of his death."[64] Panetta's explanation might justify the decision to pursue al-Awlaki, and even to kill him, but it does little to examine the ramifications of allowing the executive branch to designate U.S. citizens for assassination.

Each list serves to restrict the movement and financial transactions of its targets as part of an effort to slow global terrorism. Because each list was semipublic (meaning it could be found, but it was not easy to do so), al-Awlaki's family became aware that he had been placed on the lists. On August 30, 2010, his father, Nasser al-Awlaki, filed a lawsuit demanding Anwar's removal from the U.S. list. His concern stemmed from the fact that the U.S. list had become colloquially known as the kill list, meaning that individuals on it were specifically targeted by U.S. military and intelligence operations.[65]

While Nasser al-Awlaki's suit worked through the court system, his son remained active, particularly on social media. In October, the U.S. Congress formally requested that YouTube remove al-Awlaki's videos, many of which included calls for violence. The following month, YouTube partially complied, removing some files but leaving others in place. YouTube's parent company, Google, has always been resistant to efforts at censorship, even when the content of Internet sites is unpleasant or potentially dangerous. Google has taken a consistent stance on attempting to prevent criminal activity, particularly child pornography, but it has also made the case that freedom of speech protections must be broadly applied, even (and perhaps especially) for repugnant points of view, including the likes of al-Awlaki.

On December 7, Nasser al-Awlaki's lawsuit, filed to protect his son, was dismissed for lack of standing. Essentially, the court ruled that he had no inherent right to insert himself into a dispute between his son and the government.[66] Only Anwar al-Awlaki had the legal right to file a suit demanding that his name be removed from the kill list. Of course, given that he was in Yemen, and prohibited from air travel by his presence on the U.S. and UNSCR lists, there was virtually no possibility of al-Awlaki filing any lawsuits in person, at least without surrendering to authorities first. Anwar al-Awlaki certainly knew that he had been targeted, especially by the U.S. government, but he likely presumed his remote location in the Yemeni hinterlands offered a certain degree of immunity from direct attack.

American intelligence agencies conducted large operations in the attempt to track and neutralize al-Awlaki. This included efforts to develop human sources, interception and translation of signals intelligence, and the use of unmanned aerial vehicles to track his movements. On May 5, 2011, after months of effort, an MQ-4 Predator's sensor operator reported a potential contact. As superiors debated the veracity of the sighting, and the legality of taking direct action, someone in al-Awlaki's party noticed the aircraft. Al-Awlaki immediately fled the location and at one point changed vehicles with a pair of al Qaeda operatives while shielded by a small grove of trees. Upon emerging from the covered position, al-Awlaki's original vehicle was struck by a Hellfire missile fired from the Predator. The strike was a direct hit, killing the occupants of the car, while al-Awlaki fled in a different vehicle. Soon, it emerged that the attack had failed in its primary goal, and the hunt continued.[67]

Al-Awlaki redoubled his efforts at security in the months that followed, but he refused to enter the complete seclusion practiced by other senior leaders of al Qaeda, such as Osama bin Laden's retreat to a nondescript compound in Abbotabad, Pakistan. As such, it was likely only a matter of time before U.S. agencies again discovered his whereabouts. Yet, his commitment to secrecy kept him at large (and alive) throughout the summer of 2011. Even his teenaged son, Abdulrahman al-Awlaki, did not know where his father was hiding, beyond a vague belief that he was in the vicinity of Shawba, which was hundreds of miles from the correct location.

Like Anwar, Abdulrahman was born in the United States, and hence a U.S. citizen, although he lived with his grandparents in Yemen from age 7. He repeatedly expressed a desire to seek out and join his father, whom he had not seen for several years; but without a clue regarding Anwar's precise location, the boy could do little but hope for a future reunion. In September, he left his grandfather's home in Sana'a to look for his father, based upon a rumor of his whereabouts. Soon, he called home and let his grandparents know that he had met up with family in southern Yemen, although he had not found his father.

Anwar al-Awlaki planned to meet with Samir Khan, a fellow U.S. citizen working on behalf of al Qaeda, for breakfast, on September 30. Khan's claim to fame within the organization was the creation of *Inspire*, an al Qaeda electronic publication produced in English for Western audiences. The online magazine utilized effective visual design to captivate vulnerable audiences and encouraged young Muslims in the West to engage in lone-wolf acts of terrorism. One noteworthy edition included an article entitled "Make a Bomb in the Kitchen of Your Mom" and offered guidance on how to turn household chemicals into rudimentary explosive devices. Anyone relying upon the directions without any personal experience working with chemicals or volatile substances was more likely to kill themselves than any other targets. However, the directions, if followed correctly, would produce a potentially lethal bomb.

Unbeknownst to al-Awlaki or Khan, intelligence agents had been informed of the likely meeting and its location. As such, a Predator was already on station, ready to confirm the presence of both men. The unmanned aerial vehicle's operators fired two missiles at the car containing al-Awlaki and Khan, killing both and leaving very little forensic evidence of their presence. On September 30, President Barack Obama took a victory lap in celebration of killing al-Awlaki, stating "The death of Awlaki is a major blow to al Qaeda's most active operational affiliate. He took the lead in planning and directing efforts to murder innocent Americans . . . and he repeatedly called on individuals in the United States and around the globe to kill innocent men, women, and children to advance a murderous agenda. [The strike] is further proof that al Qaeda and its affiliates will find no safe haven anywhere in the world."[68] Everything that the president alleged about Anwar al-Awlaki is true—but the statement did not address the question of whether the U.S. government has the right to kill an American citizen without any judicial sanction or oversight.

Another major omission in Obama's statement was any mention of Abdulrahman, who was killed on October 14, 2011, while sitting at a café with a cousin and a few friends. The strike was initially explained as a targeted strike upon Ibrahim al-Banna, who was later demonstrated to have been in Egypt at the time of the strike. A second explanation claimed that Abdulrahman was a 21-year-old al Qaeda operative (a claim quickly disproven when the family provided his Denver birth certificate). After two embarrassing non-explanations, every government agency fell silent on the death of Abdulrahman, despite his family's attempts to pry loose the truth.[69] The closest thing to an answer that the White House offered came from Press Secretary Robert Gibbs, who was asked about the killing and responded: "I would suggest that you should have a far more responsible father if they are truly concerned about the well being of their children. I don't think becoming an al Qaeda jihadist terrorist is the best way to go about doing your business."[70] Thus far, the government has remained mute on the strike that killed Abdulrahman, a fact that many observers find chilling, given that it has admitted to previous mistakes in the War on Terror.

The strike that killed Anwar al-Awlaki and Samir Khan removed two very effective assets from al Qaeda, and thus it was almost certainly militarily justified. The operators took pains to avoid any collateral damage, meaning it satisfied the ethical requirement for discrimination. If al-Awlaki and Khan managed to inspire as many attacks as some reports have suggested, the airstrike that killed them was certainly proportionate—if anything, it demonstrated restraint in waiting until they could be targeted away from uninvolved bystanders. And yet, it raises the issue of whether or not U.S. citizens should be entitled to constitutional protections, including due process, before they are killed by agents of the government.[71]

The U.S. arsenal's possession of armed robotic platforms provided the only plausible means of conducting an attack upon Anwar al-Awlaki, so long as he remained in the Yemeni wilderness. Although the 2001 AUMF essentially enabled the president to conduct military activities against al Qaeda's members anywhere in the world, after 10 years in Afghanistan and 8 in Iraq, the U.S. public had tired of military adventures in distant lands. The prospect of sending troops into harm's way in Yemen, a virtually lawless state wracked by civil war for decades, was a political nonstarter, particularly for a Democratic president who had campaigned upon the promise to end the ongoing wars and bring the troops home. However, the ability to launch lethal strikes without placing any American forces in harm's way might offer a very strong temptation to any president—especially if those strikes can be launched with virtual impunity due to the enemy possessing no air defenses of any kind.

THE TEMPTATIONS OF TECHNOLOGY

The temptation to use unmanned aerial vehicles as attack platforms in the War on Terror carries some significant consequences that should not be ignored by any commander in chief. The first issue hinges upon the behavior of a population under aerial attack with no capacity to shoot down the attackers. Early air theorists argued that those populations would blame their own governments for failing to keep them safe and revolt to overthrow their government rather than absorb further punishment. However, decades of bombing campaign experience have demonstrated that such populations almost always find a different mechanism for retaliation against their attackers, rather than turning upon their own leadership. Thus, if Predator and Reaper strikes continue to kill al Qaeda and Islamic State members, those organizations will respond, but not in kind.

Normally, military operations include at least the possibility of casualties on both sides, even if they are suffered at massively disparate rates. In those circumstances, the weaker or less technologically equipped enemy might go to great lengths to directly engage the stronger opponent and pay a terrible price to do so. Yet, they will likely revel in each casualty they create. When the United States launched Operation Linebacker II in 1972, conducting the heaviest bombing of North Vietnam of the entire war, the brunt of the bombing campaign was carried by the B-52. Over the course of 11 days, the bombers inflicted enormous destruction upon dozens of targets, but they lost 16 aircraft in the process. In total, approximately 43 American aircrew died in Linebacker II, a relatively low number by most measures. More U.S. aircrew died in the single day of raids against Schweinfurt and Regensburg in 1943, for example. Yet, the North Vietnamese took great pride in each B-52 downed and recovered as many pieces as possible for display in museums and monuments throughout the country. In comparison, shooting down unmanned aircraft (or watching them crash for other reasons) has not carried the same visceral thrill for 21st-century opponents—and they have sought out other compensatory targets as a result.

Technology is not the only influence upon the strategy of a nation at war—although in the War on Terror, it seems to be a dominant consideration. In most cases, technology is only one consideration among many in the prosecution of a conflict. However, in the 21st century, the technology and the capabilities it provides seem to have become almost an end in and of themselves. When taking prisoners became awkward thanks to the scandals at Abu Ghraib and Guantanamo, the use of RPA strikes became even more prevalent, leading Jo Becker and Scott Shane to opine, "Mr. Obama has avoided the complications of detention by deciding, in effect, to take no prisoners alive."[72] Yet, this type of an approach has its pitfalls, as the case of Abdulrahman al-Awlaki demonstrated, and led David Sanger to conclude, "No American president is going to be given the unchecked power to kill without some more public airing of the rules of engagement. If the use of drones is going to be preserved as a major weapon in America's arsenal, the weapon will have to be employed selectively—and each time a public case will have to be made for why it was necessary."[73] Thus far, military robots seem to be changing the legal and moral character of war, without much public scrutiny or backlash—but the American public may quickly change its mind if these devices begin flying over American soil. As Grégoire Chamayou ominously noted, "We should not forget that when this new weapon becomes a piece of equipment for not only the military forces but also the state's police forces, it turns us into potential targets."[74] The extrajudicial targeting of American citizens certainly brings such a future closer.

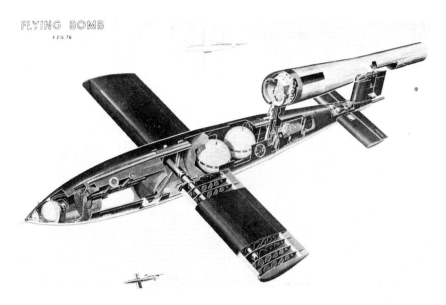

Cutaway drawing of a V-1 flying bomb, showing the complexity of the aircraft. (U.S. Air Force)

An MQ-1 Predator during a training flight. Note the AGM-114 Hellfire missile under the near wing. (U.S. Air Force; photo by Lieutenant Colonel Leslie Pratt)

An MQ-9 Reaper taxis down a runway in Afghanistan. Visible armaments include AGM-114 Hellfire missiles on outer pylons and GBU-12 Paveway II bombs on inner pylons. (U.S. Air Force; photo by Staff Sergeant Brian Ferguson)

The interior of a ground control station during an MQ-9 Reaper flight. (U.S. Air Force; photo by Airman First Class Michael Shoemaker)

A U.S. Army soldier hand-launches an RQ-11 remotely piloted aircraft. The Raven is used for tactical surveillance missions and can fly for up to 90 minutes. (U.S. Army; photo by Staff Sergeant Eboni Prince)

A PackBot EOD robot carries a stick of plastic explosives toward a suspected roadside bomb in Samarra, Iraq, on November 3, 2004. (U.S. Navy; photo by Petty Officer Class Jeremy L. Wood)

Demonstration model of a Talon SWORDS robot armed with an M16 rifle.
(Department of Defense; photo by Sergeant Lorie Jewell)

A Patriot missile battery deployed to defend Al Udeid air base, Qatar. The Patriot
has a fully autonomous option for aerial defense. (U.S. Air Force; photo by
Technical Sergeant James Hodgman)

A Phalanx Close-In Weapons System aboard the USS *Abraham Lincoln* test-fires in fully autonomous mode. (U.S. Navy; photo by Mass Communication Specialist 3rd Class Ronald A. Dallatorre)

President Barack Obama and his national security team watch a live feed from an RQ-170 Sentinel flying above the site of the raid that killed Osama bin Laden. (Official White House photo by Pete Souza)

CHAPTER 5

Robot Lawyers

So long as I hold a commission in the Army I have no views of my own
to carry out. Whatever may be the orders of my superiors, and law, I will
execute. No man can be efficient as a commander who sets his own notions
above the law and those whom he is sworn to obey. When Congress enacts
anything too odious for me to execute I will resign.

—Ulysses S. Grant, 1862

In many ways, the development of technology is primarily accomplished
for expanding control of the environment. This is certainly true of military
technology in the 21st century and has been a continual process in the U.S.
military for the entire industrial era. In a military context, control primarily involves one's ability to direct the forces within a commanded group
in the pursuit of specific objectives. If commands are what spur forces
to action, control is what places limits upon that action, and when taken
together command and control imply that a leader is trying to orchestrate
the limited use of violence in pursuit of objectives—without commands,
control simply becomes a set of stringent prohibitions on behavior, a strait-
jacket without a purpose. Without control, command becomes a compul-
sion to mob behavior, a loaded cannon with no capacity to aim.

THE DEVELOPMENT OF INTERNATIONAL LAWS
OF ARMED CONFLICT

International laws regarding armed conflict have gradually developed
over more than a century. One of the most important advances in the
laws of warfare came in 1862, with the issuance of U.S. War Department
General Orders Number 100, more commonly referred to as the Lieber
Code. The legal considerations of the Code are discussed in Chapter Five,
but it also held a significant moral dimension. Lieber, drawing upon the
work of previous scholars, philosophers, and jurists, argued that much of

the justification behind his proposed set of laws for conducting warfare derived from an immutable system of morality. His Code was primarily an attempt to link the moral urges into a legal framework, with limitations upon behavior and prescribed punishments for transgressions. In particular, Lieber noted in Article 15 that "Men who take up arms against one another in public war do not cease on this account to be moral beings, responsible to one another and to God," and thus there are limits on acceptable behavior in wartime.

The Lieber Code did not endure solely as the underpinning of later international treaties. It remained a standing order in the U.S. military until 1914, when it was superseded by *The Rules of Land Warfare* issued by the U.S. War Department. However, the latter document retained the fundamental assumptions regarding the existence of a series of moral expectations for combatants and noted, "It will be found that everything vital contained in G.O. 100 [the Lieber Code] . . . has been incorporated in this manual."[1] Perhaps the most important expression of such moral expectations came shortly after the Lieber Code, when international volunteer delegates met at Geneva, Switzerland, and formed the International Committee of the Red Cross (ICRC). The ICRC did not attempt to adjudicate international disputes; instead, it simply sought to ameliorate the conditions faced by the victims of warfare, whether combatants or civilians. By adopting this stance and scrupulously avoiding any form of condemnation of belligerents on the basis of their war aims, the ICRC developed a certain moral authority within the Western world. This moral authority, in turn, allowed it to act as a form of neutral intermediary between warring parties, a guileless observer that could effectively referee international conflicts without offering any advantages to one side or another. In this role, the ICRC served to prevent immoral activity largely by preventing misunderstandings. For example, it could inspect prisoner of war camps on both sides of a conflict and offer recommendations for how to improve conditions. It could report any transgressions it observed to all belligerents—but perhaps more importantly, it could also report acceptable captivity conditions. In practice, this supplied a powerful incentive for all belligerents to offer at least a minimal standard of shelter, provisions, and medical care for their captives, if only to ensure reciprocity for their personnel held by the other side. The ICRC presence, usually exemplified by routine and regular camp inspections, greatly reduced the mortality rate within prisoner of war camps.

The most important modern guidance regarding the behavior of belligerents is codified in the Geneva Conventions, signed in 1929 and substantially revised in 1949. One of the key provisions is a clear delineation of who qualifies as a lawful combatant and hence can claim the fundamental protections established for the safety of such combatants. In order to legally engage in conflict, an individual must satisfy four conditions:

1. He or she must bear arms openly;
2. He or she must wear a uniform, or at the very least, a distinct emblem or device recognizable at a distance;
3. He or she must be part of an organization with a hierarchical structure in which commanders are responsible for the behavior of subordinates; and
4. He or she must adhere to the laws of war.

Members of al Qaeda fail to fulfill the lawful combatants criteria for a number of reasons. Although al Qaeda fighters have been known to bear arms openly, they have far more commonly kept weapons hidden until the moment of attack. There is no recognizable uniform or symbol that is unique to al Qaeda and commonly recognized on its members. While there is some degree of hierarchy in al Qaeda's organization, it does not approach the requirement of leaders being held accountable for the behavior of their subordinates. Finally, al Qaeda certainly does not follow the laws of war—it actively targets civilians and in many ways seeks to undermine the very concept of the laws of armed conflict (LOAC).

Some organizations on the State Department's list of foreign terror organizations have taken strides toward legitimacy, in part by attempting to conform with some or all of the requirements associated with lawful belligerent status. For example, Hamas, which won the 2006 elections in Gaza and hence became its ruling authority, has moved increasingly toward compliance with LOAC. The organization, which retains its designation as a terror group according to the U.S. Department of State, has formed paramilitary organizations replete with uniforms and traditional military command structures. Even Hamas members not affiliated with its military wing tend to wear green kerchiefs as a distinguishing marker, an important step toward obtaining a certain degree of recognition and legitimacy.

Intelligence agencies do not tend to fulfill the four requirements established by LOAC to obtain lawful combatant status. In particular, their members most often do not openly bear arms or wear recognizable uniforms or symbols, as doing so would largely defeat their primary function of intelligence collection. As such, there is some question whether intelligence agents should be considered lawful combatants. Traditionally, spies have not been protected by LOAC and on many occasions have been subjected to execution upon capture. At best, they might hope to be exchanged for an enemy's spy if taken prisoner, although such exchanges have proven to be the exception rather than the norm.

Within the United States, the military is governed by the Uniform Code of Military Justice, but its requirements apply only to uniformed military personnel. It cannot be applied to intelligence agency personnel unless they are directly attached to a military unit. This has created a potential legal loophole regarding the employment of armed RPAs. The CIA has employed Predators and other armed, unmanned aircraft since their development in 2002, and, in fact, the CIA has launched a far higher number of lethal strikes from RPAs than the U.S. Air Force. This disparity is particularly obvious when examining strikes outside of Afghanistan and Iraq, where the military has been most closely engaged. RPA strikes in Pakistan, Somalia, and Yemen have primarily been launched by the CIA, which refuses to confirm or deny the number of attacks it has launched, the likely number of casualties they have caused, or the rules of engagement used when flying armed RPA missions.

COEQUAL BRANCHES IN WARTIME?

The entire U.S. system of government is designed to serve as a system of checks and balances, ensuring that no branch of government, much less an individual political leader, gains too much power over another. When the Founders designed the Constitution, they did so in part by considering both the contemporaneous governmental systems in existence (and their inherent flaws) and the historical systems they considered most ideal. Although the notion of democracy originated in ancient Greece and it sounded like a truly equitable system, the Founders realized that democracy would be impossible in practice when applied to an enormous landmass. The primitive communications available in the early republic guaranteed that any attempt to emplace a true democracy on anything more than a local level would almost certainly result in a government incapable of accomplishing anything of note. Henry David Thoreau noted "that government is best which governs least," suggesting that some might have considered such a weak central government a generally positive outcome—but responding to any form of crisis would be almost impossible.[2] Given the markedly poor function of the government created by the Articles of Confederation, most delegates at the Constitutional Convention recognized the need for at least some degree of federal government empowerment, particularly given the fact that British troops still occupied portions of the Western frontier and showed no interest in quitting their positions.

The fledgling United States could have copied the most prominent European powers and simply established a monarchy, but doing so would have repelled most citizens, given how recently the new nation had thrown off the yoke of the British royal house. Instead, the framers established a representative democracy, in particular adopting many of the concepts of the Roman Republic. However, rather than relying upon an elected Senate as the ultimate level of political power, which might have rapidly created a class of political elites, the Constitution established a bicameral legislature, in which the interests of the states and the population as a whole might be balanced.

Knowing that large legislative bodies struggle to react quickly to unexpected events, and that foreign monarchs might hesitate to treat with a constantly shifting body, the Constitution also created an executive branch, headed by a single, elected official, the president. This individual was designated to serve as the commander in chief of the nation's tiny military, in addition to serving as the chief executor of the nation's laws. Over time, this office has become associated with the ultimate civilian authority over members of the uniformed military, although selecting George Washington as the first president certainly muddied the civil-military divide for the first eight years of the office.

The third branch of the federal government, the judiciary, served to interpret the laws passed by the legislature and had the task of determining if acts by either of the other branches conflicted with the Constitution. Although somewhat less well-defined than the other branches, the judiciary clearly claimed a coequal status with the legislative and executive branches. In theory, this sharing of power between three separate entities ensured that no branch of the federal government could outstrip the others and achieve any degree of preeminence. The checks and balances of the other branches would serve to temper and restrain any such efforts.

Over the more than two centuries of federal government under the Constitution, the relative powers of each branch have risen or fallen, often according to the personalities and behavior of the officeholders. At times, the branches have directly engaged one another in a contest of wills, and none of the branches emerged untarnished from such confrontations. Consider, for example, the U.S. Supreme Court's decision in *Worcester v. Georgia* (1832). In writing the Court's opinion, Chief Justice John Marshall declared President Andrew Jackson's policy of forcible removal of Native American tribes from their ancestral lands, to be relocated west of the Mississippi River in modern Oklahoma, to be patently unconstitutional. Jackson's purported response openly defied the Court's opinion, as he stated, "John Marshall has made his decision, now let him enforce it!"[3] Given the Supreme Court's lack of any available mechanism (beyond the executive branch, of course) to enforce its decrees, it is unsurprising that the U.S. Army, commanded by Major General Winfield Scott, continued to carry out the removal policy. Thousands of Native Americans died during the forced marches, which followed a route that has since been known as the Trail of Tears due to the tribulations faced by the travelers.

Open breaches such as the one between Jackson and Marshall have fortunately been relatively rare in the nation's history. That said, the president has, upon many occasions, either ignored the demands of one of the other branches or chosen to exercise "discretion" in the enforcement of laws or court decisions. Abraham Lincoln and Franklin Delano Roosevelt, both widely regarded as being among the greatest presidents in the nation's history, each assumed powers for their office that directly contradicted congressional or judicial actions, largely in the name of wartime expedience. Essentially, both argued that the ongoing crisis of a nation at war requires the swift and decisive action of a single leader, rather than the slower and more deliberate process generally followed in times of peace. For their part, the senators of the Roman Republic recognized a similar principle and occasionally selected an individual to hold the office of "dictator," creating an all-powerful executive in times of crisis. The most well known, Lucius Quinctius Cincinnatus, assumed the dictatorship in 458 BCE to deal with an attack by the Aequi. After a rapid campaign at the head of an army composed of mass conscripts, Cincinnatus defeated the Aequi, ending the external threat and relinquished his dictatorial power. A mere 15 days after his appointment, he resumed the management of his agricultural estates, creating the model of a citizen willing to solve a crisis and then remove himself from power. Not surprisingly, many of the Founders idealized Cincinnatus as the perfect citizen-soldier, called to take up arms in an emergency by resisting the allure of ultimate power. The Society of the Cincinnati, founded in 1783, counted George Washington, Alexander Hamilton, and Horatio Gates among its original membership, clearly illustrating the principle of social leadership and responsibility for one's actions.[4]

The Constitution is extremely clear on the mechanisms required for the United States to engage in warfare. First, the legislature must vote upon a declaration of war against an enemy state, a responsibility first exercised in 1812.[5] The president, as the commander in chief of the nation's military, then holds the primary responsibility for conducting the war. At the conclusion of hostilities, the president holds the responsibility for negotiating a peace treaty, which must in turn be ratified by the Senate to bring an end to the state of war.[6] In theory, such an approach guaranteed that the nation would not lightly engage in a military conflict—such an action would essentially require the consent of the states through their representatives in the Senate.[7]

In most respects, the authors of the Constitution had a relatively realistic view of the likely future political and military conflicts faced by the United States. They did not attempt to prescribe exactly how any situation must be faced, but, in almost every way, they sought to involve at least two branches of government in every major activity. Unfortunately, they were extremely vague regarding the limits, if any, upon presidential power in times of war. Perhaps this was an effort to allow future generations to determine their own needs, or perhaps it reflected an inability to settle the issue in a fashion that would not negatively affect the chances of ratification by the states, a very practical consideration in 1787. In any event, because the Constitution offers no firm guidance regarding the limitations upon presidential power, in wartime, presidents have had a largely unrestricted hand in choosing how to conduct themselves during conflict. In turn, these decisions have established precedents regarding wartime behavior—but such precedents certainly do not carry the force of law. For example, in 1812, with the United States actively engaged in a declared war with Great Britain, President James Madison (the primary author of the Constitution), chose to hold a presidential election, knowing that he might be ejected from office if the electors disapproved of the conduct of the war. He was probably lucky the election fell in the first year of the war—had it happened in 1814, it would have occurred closely upon the heels of the British capture and destruction of Washington, D.C., and his own inept attempts at command in the field. Madison's electoral prospects differed from those of modern presidents, in that he did not have to appeal to the broad electorate—prior to the election of Andrew Jackson in 1828, the "voting public" held at best an indirect role in the selection of presidents. Yet, by holding an election, he indicated that the normal processes of government within the United States were not inherently suspended by the existence of a state of war.

In 1864, President Lincoln chose to hold elections, while attempting to quell the Confederate forces, against the advice of several of his closest confidants. His situation did not carry the force of a formal declaration of war, which might have influenced his decision, but by following Madison's example, he avoided provoking an even bigger Constitutional crisis than the one he already faced. President Franklin D. Roosevelt's 1944 election campaign also came during wartime, although it was late in World War II when most observers believed the Allied powers had almost guaranteed victory.[8] Despite his willingness to follow the Madison-Lincoln precedent of holding wartime elections, by running for reelection, he continued to break a different precedent firmly established by every previous president, that of never serving more than two terms in office. Roosevelt's landslide election, leading to his fourth inauguration, provoked some complaints that he had essentially become president-for-life. His death on April 12, 1945, technically made those complaints accurate.[9]

So, why the long discourse on the roles of each branch and the behaviors of wartime presidents? In part, the idea of presidential wartime power is important because of the lack of specific limits upon executive authority in wartime. President Lincoln suspended the writ of habeas corpus, essentially allowing him to imprison political opponents and hostile members of the press at will. Although the Supreme Court struck down his actions as unconstitutional, Lincoln largely ignored the judicial decree in favor of prosecuting the Civil War to the fullest extent possible. Roosevelt established a fairly effective press censorship institution as a means to protect critical wartime information. He also authorized massive expenditures upon unproven technology programs, the most notable of which was the Manhattan Project. Roosevelt also pursued a "Germany First" approach to World War II, even in the aftermath of the Pearl Harbor attack. He believed that, as commander in chief, he held the prerogative to determine the nation's strategy, regardless of public outcries to destroy Japan at the earliest possible moment. In this regard, Adolf Hitler's Nazi Germany became an unwitting political ally—by declaring war upon the United States on December 11, 1941, just four days after Pearl Harbor, Hitler triggered an American declaration of war against Germany in turn. Without such a declaration, which was coupled with a declaration of war against Italy on the same day for good measure, Roosevelt might have found it more difficult to focus the U.S. war effort upon Europe first. In neither the Civil War nor World War II did the president face a massive backlash from the Congress, the judiciary, or the public, and both Lincoln and Roosevelt remain extremely popular presidents in the esteem of the citizenry and historians.

By demonstrating the lack of limits upon presidential power in wartime, Lincoln and Roosevelt also established precedents for their successors in office. When President Harry S. Truman chose to commit U.S. forces to the Korean War in 1950, he did so under the auspices of the United Nations, meaning that the U.S. Senate did not declare war upon North Korea or the People's Republic of China. Even in the absence of such a declaration, however, more than 300,000 American personnel served in Korea, and 36,574 lost their lives in the war.[10] Truman certainly was not the first president to conduct a war without a formal declaration. Even ignoring more than a century of conflict between the federal government and dozens of Native American tribes (a situation that overlapped almost every presidential administration of the 19th century), fighting wars without legislative permission has a long tradition in the United States. The second president, John Adams, fought a naval conflict with France from 1798 to 1800, dubbed the Quasi-War. His successor, Thomas Jefferson, authorized a naval war against the Barbary states of North Africa, culminating in a land invasion of Tripoli by a handful of U.S. Marines, backed by 500 mercenaries. In the 20th century, President William McKinley oversaw the Philippine War (an outgrowth of the declared Spanish-American War of 1898). This conflict stretched into President Theodore Roosevelt's administration and formally continued until 1907. McKinley also authorized U.S. participation in a multinational invasion of China in 1900 to help quell the Boxer Rebellion. In 1914, President Woodrow Wilson sent U.S. forces across the Mexican border in pursuit of Pancho Villa, allowed an amphibious invasion and occupation of Veracruz, and committed U.S. troops to a coalition invasion of northern Russia in an ill-fated attempt to overturn the Bolshevik victory in the Russian civil war. Under Warren G. Harding and Calvin Coolidge, American forces invaded and occupied a host of Latin American nations, activities that certainly qualified as acts of war under international law, regardless of the lack of a domestic declaration of such. In short, presidents have often engaged in belligerent behavior, even if their particular circumstances did not merit a declaration of war by the Senate.

The most significant example of a president conducting a war without a formal declaration of war spanned the administrations of five individuals. President Dwight Eisenhower committed U.S. troops to an advisory mission in South Vietnam after French colonial forces were expelled from Indo-China. John F. Kennedy transformed the U.S. role in South Vietnam from an advisory capacity to active engagement in hostilities. After Kennedy's assassination, Lyndon B. Johnson massively expanded the war effort, with the number of U.S. troops in Vietnam peaking at 525,000 in 1968. His successor, Richard Nixon, claimed to have a secret plan to win the war during his first election campaign, yet when he stood for reelection in 1972, the war still raged. By the time the last U.S. military forces evacuated the country, Nixon had resigned the office of the presidency in favor of Gerald Ford, and the unsuccessful U.S. effort had cost the lives of 58,220 American service personnel.[11]

THE WAR POWERS RESOLUTION

In 1972, after nearly a decade of U.S. involvement and with no end to the Vietnam War in sight, congressional candidates seeking election launched a bid to assert legislative authority and end the U.S. involvement in the Vietnam War, regardless of the consequences a unilateral withdrawal might create. The Congress elected in 1972 included dozens of new members, most of whom vowed to eliminate any federal funding for the war in Vietnam and to explicitly prohibit the Department of Defense from expending any further resources upon the war without congressional approval. Only the minimum resources necessary for the evacuation of U.S. personnel could be expected as soon as the new Congress took office, and it was unlikely that even a presidential veto would suffice to keep Congress from asserting its authority. Anticipating the problems this policy might create, the Nixon administration redoubled its air campaign against North Vietnam, launching Operation Linebacker II. This campaign involved the first use of the heaviest bomber aircraft in the U.S. inventory, the B-52, against targets in North Vietnam. While it could not singlehandedly win the war, Linebacker II certainly created an enormous amount of pressure upon the Vietnamese to conclude armistice negotiations, resulting in the short-lived Paris Peace Accords.

The Vietnam experience amply demonstrated to Congress that an unchecked presidency could conduct ruinously expensive wars on its own initiative. To reassert the key role of Congress in deciding when and how U.S. military forces may be utilized in international conflicts, the House and Senate passed the War Powers Resolution (WPR) in 1973, overriding President Richard Nixon's veto. Essentially, the WPR prohibits the president from deploying forces for more than 60 days without explicit approval from Congress. Further, the president is required to notify congressional leaders within 48 hours of any such deployment. These provisions recognize that the president might need to immediately react to a crisis without seeking congressional approval but that once the initial emergency situation has stabilized the approval of at least two branches of government will be integral to continuing the engagement in international conflicts. In theory, this reasserts the traditional congressional role, but, in practice, presidents have not always followed the explicit requirements of the WPR. Most presidents since Nixon, regardless of party, have expressed some trepidation regarding the WPR, which they tend to perceive as an unconstitutional curtailment of the president's powers as commander in chief.

U.S. presidents since 1973 have not always followed the letter or the spirit of the WPR, although no president has attempted to conduct a full-scale war without congressional approval. Presidents are keenly aware of any limitations placed upon their authority as commander in chief, and for most of the nation's history, wartime presidents have worked to expand the power of their office by pushing the boundaries of acceptable presidential conduct during conflicts, setting precedents that if unchallenged might become assumed powers. Few presidents have shown any hesitation to engage in this behavior, regardless of the fact that any power thus gained will be conveyed to all future presidents, some of whom may not have the same ideological beliefs as the current occupant of the White House. In this regard, presidents of all political parties and persuasions have conducted themselves almost uniformly in harmony.[12]

One of the primary reasons for creating the WPR was to prevent the president from dragging the nation into a war without legislative approval, which in theory involves some discussions with the citizenry and hopefully includes some consultations on strategy and objectives. However, the WPR also solved an emerging dilemma within political circles about how to conduct a war without formally declaring the same. The UN Charter, signed in 1945, was in large part designed to prevent future international conflicts. Article 2 requires that "All Members . . . settle their international disputes by peaceful means in such a manner that international peace and security, and justice, are not endangered."[13]

For all intents and purposes, the UN Charter essentially forbids all but defensive wars and limits how one might define such a defensive struggle.[14] In lay terms, the United States would almost certainly have to be directly attacked by a foreign nation before it had a legal right to declare war. Given the U.S. primacy in global military power, particularly since the early 1990s, such an overt act has been extremely unlikely since at least 1945.[15] Of course, the United States is free to provide assistance to other member states that have been attacked, a provision that has been used to justify a number of conflicts undertaken without UN sanction. Otherwise, though, the UN Security Council must authorize military attacks, and, because the five permanent members of the council (China, Great Britain, France, Russia, and the United States) can each veto any military authorization, a consensus regarding the use of force is necessary among the five permanent members before it will pass.

The massive expansion of military robotics offers a unique challenge to the WPR. The WPR explicitly requires that the president provide written notification to Congress within 48 hours of ordering U.S. troops to a foreign country, except in cases of normal troop rotation and resupply.[16] However, military robots, while capable of engaging in acts of war, by definition, are not soldiers, sailors, airmen, or Marines. So long as they are controlled remotely, or function entirely autonomously, they cannot be regarded as expressly covered by the WPR, particularly if they are flown by crews in the United States. Although some might argue that the use of military robots in lieu of uniformed humans might violate the *spirit* of the WPR, they can hardly be equated with placing U.S. personnel in harm's way. Thus, the continued expansion of military robotics creates an unanticipated loophole in the WPR—one which three presidents (George W. Bush, Barack Obama, and Donald Trump) have exploited, whether intentionally or not. So long as warfare is conducted entirely via remote platforms, it is theoretically possible for a president to engage in warfare without any congressional oversight or approval.

The advance of remotely operated military robots and the development of cyber weapons might obviate the rationale behind the WPR. After all, when the WPR was crafted, there had been very few attempts to use remotely operated or autonomous weapons, and there was no such thing as a cyber domain—only a handful of computers had been linked in the distant precursor to the Internet. Thus, the creators of the WPR can hardly be faulted for failing to envision how technology might inherently alter the definition of "armed forces" and even the question of whether they can be considered to be "in action" if the only human actors engaged in conflict remain safely ensconced on American soil, immune from any direct action by enemy forces. Further insulating the president, many of the cyber operators and remote-aircraft pilots are technically civilian contractors who are not part of the uniformed military (or exactly members of the intelligence agencies, for that matter). Thus, the development of these new systems might allow the president to engage in warfare without formally notifying Congress, much less requesting its authorization. After all, if a software program fails to meet its intended goal or a remotely piloted aircraft is shot down during its mission, no American lives, military or otherwise, can be considered to have ever been at risk. If no U.S. lives were at risk, Congress has a very weak argument that the president has broken any laws by not providing formal notification, and the public is unlikely to be incensed if the president chooses to attack foreign citizens in a distant land without using U.S. troops. This is especially true if those citizens can be plausibly (or even implausibly) tied to a terror organization that has launched attacks against U.S. interests, possibly even attacks on U.S. soil.

The principle that the WPR might not apply to remotely operated platforms has been tested by the two most recent presidential administrations, and the precedents that each set will likely be upheld by subsequent commanders in chief, regardless of party affiliation. In 2011, the Obama administration chose to provide assistance to Libyan insurgents in the form of remotely piloted aircraft assets and aerial refueling for other NATO members engaged in aerial activities. The unmanned aircraft being flown over Libya included some armed variants, including MQ-1 Predators and MQ-9 Reapers, each of which fired upon Libyan government targets during the campaign. President Obama not only failed to provide formal notification of the actions to Congress but also steadfastly refused to do so on the grounds that they were not required by the WPR.[17] In April 2017, President Trump launched several dozen cruise missiles against Syrian military targets in retaliation for the Syrian government's use of chemical weapons. Trump's administration notified Congress of the action, in keeping with the WPR, but his actions might have violated the Constitutional provision that Congress has the prerogative to declare war. His attacks were not launched in self-defense, nor were they in support of an aggrieved ally, meaning they were certainly an act of war under international law.[18]

ASSASSINATIONS AND THE U.S. MILITARY

In 1976, President Gerald Ford signed Executive Order 11905, formally prohibiting members of U.S. intelligence agencies from engaging in assassinations. Presidents Jimmy Carter and Ronald Reagan renewed the ban through executive orders. Each iteration of the ban included some key verbiage to illustrate the purpose behind the ban and the expectations for intelligence personnel. The original order included the statement "No employee of the United States Government shall engage in, or conspire to engage in, political assassination." When President Carter renewed the ban, it included the same prohibition, without the adjective "political" in the statement. President Reagan maintained the same phrasing as Carter in Executive Order 12333, but like his predecessors, failed to actually define "assassination," leaving open the question of whether terrorists might be attacked as a part of national self-defense. In general, the presumption has been that wartime killing should not be considered assassination, even if it targets a head of state or major political figure, as there is no element of surprise in an existing conflict.[19]

In 2002, well before unmanned aerial vehicle airstrikes became commonplace in the effort to hunt down and kill al Qaeda leaders, Bruce Berkowitz illustrated three situations in which the U.S. government might be allowed, through its agents, to engage in killing for political purposes. The first is during a law enforcement operation in which an individual places others at immediate risk. The second is during military operations. The third is after an individual is sentenced to death for a capital crime. The War on Terror certainly contains some elements of each of these concepts—the law enforcement aspect of apprehending the individuals responsible for the September 11 attacks, the military activities necessary to capture or kill enemy combatants as a means of protecting the nation from further attacks, and the punishment of those tried in absentia for their crimes are all potential justifications for the government to engage in targeted killings.[20]

While the U.S. military has its own intelligence-collection apparatus, most military intelligence has focused primarily upon the military capabilities of enemy and potential enemy nations. As such, Department of Defense agencies had far less practical experience collecting intelligence against non-state actors such as al Qaeda. The CIA, on the other hand, had long focused upon al Qaeda and other extremist groups. In the aftermath of September 11, the CIA naturally took the lead intelligence role in the fight against al Qaeda, to include obtaining its own fleet of RPAs for both ISR and lethal strikes. According to Brian Glyn Williams, "CIA head George Tenet had serious qualms about the new killing technology and the ethics and legality behind its use. The consensus in the CIA was that 'aircraft firing weapons was the province of the military.' According to one former intelligence officer, 'There was also a lot of reluctance at Langley to get into a lethal program like this.'"[21] And yet, once the utility of firing upon a target almost immediately after spotting it became obvious, the CIA embraced the new technology. Williams continues, "At a relatively early stage of the game, the CIA came to see the drones as an advantage and a liability. They were an unprecedentedly accurate tool for killing the likes of al Qaeda number three Muhammad Atef 'al Kumandan,' but they were still reliant on solid humint [human intelligence sources] in order to be effective."[22] It is illustrative that there was very little hand-wringing about whether the CIA should be in the business of killing in a war zone—only the mechanism (an obviously military aircraft) caused the organizational leaders any pause.

Military personnel and intelligence operatives are governed by different legal codes, both in domestic and international law. Within the U.S. Code, the Department of Defense is governed by Title 10, an enormous body of law that establishes the uniformed military as the primary organization tasked with national defense and offensive violent actions on behalf of the nation. The national intelligence agencies are codified in Title

50 and for the most part are limited to the collection of intelligence rather than engaging in lethal operations. When an attack can be construed as an act of war rather than an assassination, despite creating the same effect, it becomes a legal application of violence. Thus,

The oddity of labeling what appears to be good old-fashioned assassination as an act of *war* is that it seems to change nothing about the homicide itself. It serves the purpose instead of making collateral damage possible where it was not before. The irony, then, is that more people, even known to be innocent, become subject to slaughter when an act of assassination is rebranded as an act of war, despite the fact that the other distinctive features of warfare are conspicuously absent.[23]

When President George W. Bush pronounced the existence of a global War on Terror, he was referring to the global reach of al Qaeda and the fact that it had established terror cells around the world. Although the military might be called upon to conduct large-scale invasions of Afghanistan and Iraq, and anywhere else that might offer sanctuary to al Qaeda, the struggle with the terror organization itself would not take on all of the aspects of a conventional conflict and would probably spread to dozens of locations. Yet, committing overt military force to action across international borders carries with it major consequences, particularly if it is done without the permission of the affected countries. Thus far,

The American claim has been that it has the legal option to use drones against targets that pose a threat if the foreign government is unwilling or unable to take action on its own to remove the threat, with the underlying legal presupposition being that a government has an obligation not to allow its territory to be used as a launching pad for transnational violence. What becomes clear, however, is that both the globalizing of conflict, and of threats and responses, are incompatible with a state-centric structure of law. If a legal order is to persist under these conditions, it must be globalized, as well, but there is an insufficient political will to establish and empower global procedures and institutions with such effective authority. As a result, the only alternatives seem to be an inchoate imperial regime of the sort that presently prevails, or an explicit global logic of reciprocity. This has not yet happened.[24]

At least in theory, the State Department needs to liaise with local authorities in the foreign country to obtain permission for a military strike. Doing so in many locations all but guarantees that the targets of the strike will be alerted to the impending action. If the United States does not notify the local authorities of its intentions, the use of military force legally constitutes an act of war and certainly violates a host of international laws. Members of both houses of Congress must be notified of the deployment of troops, an action that all but guarantees it will be revealed to the press and hence the American public in short order. There is no such provision regarding the activities of intelligence operatives, even if they include lethal attacks. Therefore, there are a number of compelling—though legally questionable—reasons why using anything other than military units for attacks against terrorist groups is an attractive option.

U.S. PRESIDENTS AS COMMANDERS IN CHIEF

U.S. presidents have always been designated as the commander in chief of military forces, but only two, George Washington and James Madison, literally stood at the head of an army, commanding military forces in the field while they were in office. Several others held significant military experience prior to their election, and this undoubtedly influenced their relationship with active-duty personnel during their administrations. In their role as commander in chief, presidents have set the objectives for military action and usually heavily influenced the development of strategy in pursuit of those objectives. But once the military is set in motion, controlling its activity becomes a far more difficult problem, particularly when confronted with the chaos and unpredictability of warfare. Prussian military theorist Carl von Clausewitz understood the difficulty of controlling military forces even in the immediate environment of the battlefield—he considered political leadership to be at least partially irrelevant once the shooting began. But U.S. presidents have always proven loathe to risking their military forces and have striven to mitigate that risk by exerting personal control over their activities when possible. Technological innovation, particularly as it pertains to the transmission of information, has proven the great enabler for efforts toward military control.

President Washington knew from personal experience during the Revolutionary War just how hard it can be to control a field army, particularly one made up of volunteers who considered it their inherent right to question almost every aspect of military activity. Throughout the Revolutionary War, he sought to maintain at least some degree of control over his military forces, with limited success, though his army neared mutiny on several occasions. Eventually, Washington was forced to entrust subordinates to command geographically separated forces without anything more than broad strategic guidance, because there was simply no means of rapid and secure communication that would allow him to exert any degree of control. In the third year of his presidency, Washington decided to personally lead the march into Pennsylvania to put down a budding rebellion, and his presence at the head of the army might have sufficed to cool the rebels' ardor. Subsequent presidents have struggled a great deal more in their efforts to command and control the U.S. military, even with their inherent right to do so ensconced in the Constitution.

When President James Madison attempted to oversee the War of 1812 from Washington, he did so knowing that he could exert control only over the forces within his immediate vicinity. Communication with the forces arrayed on the New York border with Canada required a minimum of one week on horseback in each direction, weather permitting. News of the fighting near Detroit did not reach Washington for weeks, and the British invasions of the Gulf Coast cities were long decided by the time Madison knew they had occurred. He needed to impart his overarching vision to his generals and then trust them to carry it out—something they utterly failed to accomplish. General William Hull surrendered the key outpost at Detroit almost without firing a shot; General Stephen Van Rensselaer led a disastrous invasion into Canada but half his forces refused to cross the border to support him; and General Andrew Jackson conducted a private war against the Creek Indians rather than trying to repel the British. Eventually, Madison assumed personal command of the remaining forces near the nation's capital and found that exerting control over an army is a far different matter than merely commanding it. What might Madison have done with modern communication systems? If his leadership in the aftermath of the burning of Washington is any indicator, it likely would have allowed him to squander his military resources even faster![25]

During the Mexican War, President James K. Polk had little choice about trusting his field commanders to conduct war on behalf of the nation. Short of departing Washington to personally inspect the front, Polk had to wait for weeks for news of the invasion of Mexico. Yet, Polk feared that if he entrusted too much to a single field commander, he might create a national hero in the form of a victorious general, and all of the Army's highest-ranked officers happened to be ardent supporters of the Whig Party, while Polk was a Democrat who had vowed not to seek reelection. Thus, Polk put Zachary Taylor in charge of the invasion of northern Mexico but stripped half of his army away to assist Winfield Scott's march on Mexico City. Ironically, though it was Scott who eventually completed the victory and reaped much of the military glory, Taylor succeeded Polk in the White House.

While the U.S. Civil War raged, President Abraham Lincoln struggled to find a field commander to his liking. The first two years in the eastern theater saw a continual churn in the command structure, with multiple generals in chief and commanders of the Army of the Potomac. In comparison, the Confederates named a commander of the Army of Northern Virginia, General Joseph E. Johnston and stuck with him despite the fact that Confederate president Jefferson Davis had loathed him since their days together at West Point, and neither could find the courage to put aside their enmity.[26] Johnston remained in command until he was shot in the chest, his successor and West Point classmate, General Robert E. Lee, assumed and remained in command until the Confederate surrender at Appomattox. Lincoln's frustrations often manifested through direct communications with his field commanders, made possible via the newly invented telegraph. Unlike previous U.S. presidents, Lincoln could expect to learn news of the front within a few hours of major battles—and he spent many nights in the telegraph office, waiting for developments from hundreds of miles away. While this allowed Lincoln to engage in more effective command, because he could normally communicate with most of his field commanders and coordinate their efforts, it also offered the temptation of trying to exert control over the fielded forces. On a number of occasions, Lincoln's frustration with the lack of progress clearly boiled over and was communicated in sharply worded messages instructing his generals to get their forces moving toward Richmond. On one occasion, Lincoln remarked, "If General McClellan does not want to use the army, I would like to borrow it for some time."[27]

When telephones were invented, they offered a somewhat expanded communication ability over the telegraph, although they were fairly undependable at first, and of course required a wired connection before the two parties could speak to one another. President Rutherford B. Hayes ordered the first telephone installation in the White House, in 1877, but it had only a single connection, to the Treasury Department. As a result, a connection to the White House telephone was established from the Treasury Department simply by dialing "1" and awaiting a response. Not surprisingly, Hayes received few calls over the course of his administration. The telephone service of the United States expanded substantially over the following 20 years but not enough for President William McKinley to use telephone communications to direct the Spanish-American War—for the most part, he supplied broad guidance to field commanders through the War and Navy Departments and then awaited the results.

The development of radio sped communications even further, but it came with a few caveats. First, its range, while theoretically unlimited, had practical constraints created by the line-of-sight nature of radio waves passing from one receiver to another—radio did not suddenly allow communication around the globe. Second, radio waves are emitted by signal towers in every direction, not solely upon a narrow predefined path, as was the case with telegraph signals traveling along a single wire. This made radio waves far more vulnerable to interception, if one could tune in to a particular frequency. Although coded signals soon became the norm, they also slowed down communications a great deal. Thus, the development of radio did not trigger President Woodrow Wilson to attempt the micromanagement of World War I, although he did personally negotiate the treaty to end the war. Likewise, in World War II, Presidents Franklin D. Roosevelt and Harry S. Truman did not insist on controlling anything more than the broad arcs of American strategy, although both reserved the right to make the most consequential decisions that might have major political ramifications within the Allied alliance. Truman continued this pattern in the Korean War—some historians have argued that he gave

General Douglas MacArthur too much leeway and then overcompensated by removing him from command.

One of the most prominent examples of presidential micromanagement came during the Vietnam War. In 1965, the United States commenced Operation Rolling Thunder, a coordinated industrial bombing campaign of North Vietnam modeled after the World War II strategic aerial attacks. President Lyndon Baines Johnson, who fancied himself a gifted strategist, soon began to demand personal approval of the aerial targeting list. He also held weekly strategy lunches, to which his highest-ranking military officers were deliberately not invited.[28] This intentional slight reminded the military of the command relationship with the presidency but also suggested that the president could lead a war as a part-time endeavor, a topic to be picked up for a few hours each week but then left to the military professionals for the rest of the time. Not only was the micromanagement frustrating for the military, it was also counterproductive, as militarily significant targets were frequently bypassed in favor of political maneuvering and bombing halts were called whenever North Vietnam indicated it might be willing to engage in diplomacy. By attempting to tightly control the forces under his command, Johnson became an incredibly ineffective commander. He demanded enormous volumes of information, including raw photo reconnaissance stills taken by repurposed targeting drones, but then refused to heed the expertise of military professionals who attempted to explain the significance of what they showed.

During the Vietnam War, manned aircraft initially conducted the dangerous photoreconnaissance runs, flying over heavily defended sites armed with nothing more than cameras. The shift to unmanned platforms mitigated some of the risk but also came with a less efficient performance, as the unmanned photo drones flew entirely preplanned routes and could not react to unexpected circumstances. Their film canisters had to be manually recovered and developed, although near the end of the conflict, some of the photo drones were equipped with television cameras that allowed a direct broadcast of imagery.[29] The ideas tested out in Vietnam became more fleshed out in subsequent decades, such that by the time of Operation Enduring Freedom, a Predator feed had been connected directly to Langley, with a subsidiary connection soon added at the White House. President George W. Bush could watch real-time activity over the battlefields of Afghanistan and, if he chose to do so, could communicate directly with the crew flying the Predator. Naturally, he could also react to the images displayed on the video feed—and he reportedly contacted field commanders on a frequent basis to inquire about certain aspects of the conflict and the imagery it produced.

The new presidential tendency to examine a war at the tactical level was nowhere more on display than in the photos that emerged in the aftermath of the raid that killed Osama bin Laden. It was soon clear that President Barack Obama and the key members of his national security team sat in the White House Situation Room and watched the raid in real time as it was broadcast from cameras mounted on an RQ-170 Sentinel circling overhead.[30] Thus far, there is no evidence that any of the political observers sought to interfere with the actions of the attacking team, but there is nothing to stop a president from choosing to communicate directly with a fielded special operations team on a future raid of strategic importance.

The first Predator feed required a significant portion of the existing satellite bandwidth capacity of U.S. communications. During Operation Desert Storm, the entire satellite communications network for U.S. forces in-theater amounted to 100 Mbps. In the later invasion of Iraq, the bandwidth requirements jumped to 2,400 Mbps in 2003 and continued to expand exponentially throughout the conflict. To put things in perspective, a single RQ-4 Global Hawk requires 500 Mbps to transmit its sensor feed, meaning that one modern sensor platform uses five times as much satellite bandwidth as the entire military force deployed in 1991.[31] The constant demand for expansion paralleled the ever-increasing demands for information—as the situation on the ground deteriorated, Army commanders seemed almost consumed with the attempt to create a perfect information situation. For its part, "The Air Force was not interested in commanding such an awkward, unproven weapon [Predators]. Air Force doctrine and experience argued for the use of fully tested bombers and cruise missiles, even when the targets were lone terrorists. The Air Force was not yet ready to begin flying or commanding remote control planes."[32] Air Force culture simply had no mechanism to allow for unmanned aircraft, particularly armed variants.

Of course, all the information in the world is almost useless if a commander does not have the capacity to act upon it. In Iraq, during the height of the insurgency, commanders could see and identify a host of fundamental problems—but found that they had little capacity to actually solve the underlying problems that could be addressed, which explains General Peter Chiarelli's focus upon SWET (sewage, water, electricity, and trash).[33] These elements were visible aspects of urban life that might be improved through the application of technical problem solving. Chiarelli's theory was akin to the broken-window policing concept. Its proponents argue that police should focus upon changing the physical environment in which crime thrives—rather than ignoring the petty vandalism of a broken window or a bit of graffiti, they should report it and strive to get it fixed on the grounds that a community can insulate itself against crime if the inhabitants unite in their intolerance of criminal activity. Chiarelli confronted the exact problem foreseen by General Shinseki in the planning stages of the war in Iraq: there were simply too few U.S. troops to provide basic security, the first element of an occupation, and no amount of technical sophistication could upend that situation.

THE DIFFICULTIES OF DECLARING WAR

Declarations of war have been relatively rare in U.S. history, and there has not been one for more than seven decades. Not since the immediate aftermath of the Pearl Harbor attack has the United States served notice to another nation that it would engage in formal hostilities. In theory, the United States renounced war as an instrument of policy in 1928 with the signing of the Kellogg-Briand Pact. And yet, there was little protest that the interwar agreement should prevent the United States from engaging in World War II, either from domestic or from international sources. First, several signatories to the pact, notably Britain, France, and Germany, had already been fighting since 1939, suggesting that the pact was no longer binding in fact, regardless of the question of whether it still applied in theory. Second, the national mood was such that there was simply no possibility of absorbing the sneak attack on the Pacific Fleet without striking back. Anyone in the United States arguing that Japan had a legal or moral right to attack Pearl Harbor was far more likely to receive a punch in the mouth than a standing ovation. In the same fashion, when four hijacked jetliners crashed and killed nearly 3,000 civilians, there was no possibility that the United States would fail to respond or even limit itself to the previous types of retaliatory attacks that had so obviously failed to dissuade al Qaeda from further attacks. Earlier cruise missile strikes in Sudan had, if anything, only encouraged al Qaeda to continue its activities, on the assumption that the United States lacked the will to conduct a close-quarters fight using heavy military forces. If the response to over 2,000 dead at Pearl Harbor was to declare a war against Japan, the response 60 years later to 3,000 dead in New York, Washington, and Pennsylvania was essentially guaranteed to mirror the earlier situation.

When the United States went to war on the Korean Peninsula in 1950, it did so under the auspices of a UN unified command. The UN Charter forbids member states from launching aggressive wars against one another, but it certainly does not prohibit fighting as an aggrieved victim. In fact, it suggests that the civilized nations of the world have a legal and moral duty to come to the aid of the victim nation that has been struck by an act of war. Barring a veto from one of the permanent members of the UN Security Council, any nation that is willing and able can join in the collective effort. Although American commanders formally led the effort in Korea, they did so as the head of UN Command, while retaining their own position in the U.S. Army. In a similar fashion, when member states of the North Atlantic Treaty Organization have collectively chosen to go to war, the formal commander has usually but not always been an American.

THE PROBLEMS OF DECLARING WAR
ON TERRORISM

There are multiple barriers to simply declaring war against al Qaeda. The first, and most obvious, is that al Qaeda is not a state, and thus the centuries-old practice of interstate warfare, to include formal declarations, hardly applies. While al Qaeda comprises a tangible group and is thus more susceptible to the basic application of violence common to making war than broad concepts like poverty or drugs, international law does not really admit the possibility of a non-state party to war. In fact, if anything, the laws of war discourage such a designation, if only because no non-state actors have ever been invited to sign the governing international laws of armed conflict. Further, to allow a declaration of war against a non-state actor would be to essentially recognize that actor as a lawful belligerent that could theoretically claim the protections of the laws of armed conflict. Another complicating factor is the question of how wars end—traditionally, they end with either a peace treaty or the unconditional surrender of one of the belligerents. However, in the case of al Qaeda, there is no recognized body for the group that could legally sign a peace treaty, even if it chose to do so. Also, the United States has a firm precedent of not engaging in negotiations with terrorists, despite a few instances to the contrary, and would be unlikely to raise al Qaeda to the level of cobelligerent equal through such discussions. Finally, there is the key question of whether al Qaeda and its members could ever be trusted to live up to the provisions of any agreement terminating the conflict or whether the only means to end the threat presented by al Qaeda is the complete eradication of the group. If annihilation is the only mechanism to end the conflict, there is no compelling reason to declare a state of war, with its resulting domestic effects and requirements, as there will be no mechanism to end the war if there is no enemy left to negotiate.

For its part, while al Qaeda might find it convenient at times to demand that the United States live up to all of the provisions of the Geneva Conventions and other key international rules regarding warfare, in reality, al Qaeda's fundamental goal is to pull down the current international order. From the perspective of non-state entities like al Qaeda, the current laws of war, which were largely devised by the victorious powers of World War II and imposed upon the rest of the world, are largely devised to ensure that the most powerful states of 1945 retain their preeminence within the global political structure. Examining the makeup of the UN Security Council, and in particular the permanent membership upon it, illustrates the point quite effectively. On two occasions, permanent seats on the council have changed control. When the Chinese civil war ended in 1949, the Chinese Communist Party assumed control over mainland China, but the previous ruling party, the Kuomintang, fled to Taiwan and established a government in exile. For more than two decades, the proper holder of the Chinese seat remained under debate, such that there were effectively only four permanent members of the council. In 1971, President Nixon offered to drop U.S. opposition to transferring the seat to the communists in exchange for the end of Chinese intervention in Vietnam. The other transfer occurred in 1991, when the Soviet Union collapsed and its seat transferred to the newly formed Russian Federation. If a neutral observer sought to redesign the Security Council in the 21st century and wished to retain five permanent members, it would be difficult to do so in a fair manner that resulted in the same membership.

If membership on the Security Council was limited to the five largest economic powers on the globe, France and Russia would lose their seats to Japan and Germany, a marked shift from the composition created at the end of World War II. If population size became the premise for membership, the new council would include India, Indonesia, and Brazil, while retaining only the United States and China. A selection based upon total landmass would induct Canada and Brazil in place of Britain and France. If one chose to design the council on a regional basis, it would be difficult to retain three European powers, or two English-speaking nations, or a council entirely comprised of nations in the Northern Hemisphere. Perhaps membership on the council would be based upon military might (a much more difficult measure than population size or economic production), but even so, retaining Britain and France would be relatively difficult to justify. Perhaps membership should be extended to nuclear nations, which would allow for the admission of India and Pakistan. Of course, this option would also require an invitation to North Korea, might require the admission of Israel if it ended its policy of ambiguity, and would incentivize nations that wished to pursue nuclear weapons.

Declaring war upon a behavior, such as terrorism, creates its own fundamental problems, not least of which is the inability of the international community to settle upon a single definition of "terrorism." The old adage "one man's terrorist is another man's freedom fighter" certainly applies in this case. Many of the most revered figures of the American Revolutionary War might well have been labeled terrorists by their British adversaries had the term actually been in use at that point in history. If a state declares war on an activity such as terrorism, how can one know that the war has been won? There is no prescribed period of time that might serve to determine that terrorism had been annihilated, and a single terror attack might restart the clock at any point. On a number of occasions, powerful maritime empires have effectively declared the seas to be free of piracy, sometimes for periods lasting centuries, yet pirates in the waters off East Africa and in the Straits of Jakarta both demonstrate that an action can never be truly eradicated; it can only be suppressed for a period of time. This problem is akin to the problems associated with the overuse of the term "war," particularly in political shorthand. In many ways, it is equally silly to declare war on drugs, poverty, or Christmas—all of these "wars" are anything but a war and muddy the discussion of international conflict.

The third problem with a formal declaration of war upon al Qaeda or similar groups, should the first two not be enough to prohibit such an action, is one that has accompanied every declared war in U.S. history. The Constitution is very clear that declaring war is an act of the legislature, and it leaves no ambiguity that the president serves as the commander in chief of the nation's military force. The Constitution is not particularly forthcoming on what the limits of presidential authority in wartime might be, but past practices have suggested that there are few limits to the power of a determined wartime president. Abraham Lincoln suspended the writ of habeas corpus to imprison political enemies, forced states to contribute troops to fight the Civil War, and even debated cancelling the 1864 elections. Franklin Delano Roosevelt ordered unprecedented amounts of deficit spending, gave away billions of dollars in military hardware, ordered the internment of hundreds of thousands of U.S. citizens on the presumption that they might contemplate sabotage, and kept major weapons projects secret from virtually everyone in the government, including his own vice president. Both Lincoln and Roosevelt are widely regarded as excellent presidents, in part for their wartime leadership. One key element that all wartime presidents to date have had in common is a willingness to build upon the powers assumed by predecessors. This is no surprise: many scholars have identified a gradual rise in the so-called imperial presidency, and the relative power of the office has definitely grown when compared to that of the other branches of government.

When the United States declares war, the president essentially gains the powers of a dictator, at least on a temporary basis. In a time of national crisis, particularly a war of national survival, this is the only practical means to fight a conflict while maintaining a democratic system of government. However, there is also only one way spelled out in the Constitution for ending a formal state of war, and that is for the president to negotiate a treaty that is subsequently ratified by the Senate. In some conflicts, this method of ending the war has proven problematic. During the Mexican War, U.S. troops invaded and occupied Mexico City and effectively disbanded or destroyed the Mexican government. Until the U.S. occupation forces essentially demanded the formation of a new Mexican government, there was no mechanism to negotiate an end to the conflict. At the end of World War I, President Woodrow Wilson directly participated in the treaty negotiations at Versailles but then found that he could not get the treaty passed by a hostile Senate. As a result, the state of war between the United States and Germany remained in effect until 1921, when the nations signed a peace separate from the rest of the belligerents. If there is no means to determine victory in a war against an idea or a behavior, there is also likely no formal enemy with whom to negotiate an end to the conflict. If Congress offered a full declaration of war against terrorism, the United States might thus find itself in a state of perpetual war with no legal means to end it, or it might have a president with little or no interest in returning the wartime powers back to the other branches of government.

THE RISE OF LETHAL MILITARY CONTRACTORS

Military contractors have emerged as a ubiquitous aspect of U.S. operations. In part, this is because it is often cheaper to hire a specific capability on a short-term basis than it is to develop and maintain such a capability on a semipermanent basis. Legally, contractors allow augmentation of the size of a military force while remaining under the statutory limitations on the number of personnel within each service. Another legal aspect of military contractors that often goes overlooked is that they provide a certain degree of insularity between a military commander and events on the battlefield. Military contractors are modern mercenaries—they augment armed forces in exchange for fiscal gain. But, as with traditional military forces, they might be called upon to engage in activities that could cause uniformed troops to balk. In the 21st century, that includes the employment of remotely piloted aircraft in the hunt for terror subjects.

Matt J. Martin reported that in the early days of Predator strikes, there was a massive problem with intelligence "stove-piping."[34] In particular, Predator pilots and sensor operators became frustrated with the number of permissions required to fire a weapon. In his perspective, "The U.S. military went to superhuman lengths to avoid civilian casualties. The Iraqis took advantage of our restraint and painted warnings on the roofs of strategic installations: *Contains Human Shields.*"[35] The CIA, which is governed by a completely different section of the U.S. legal code, presumably does not follow the same procedures as the military regarding the use of armed Predator strikes. It is impossible to determine from materials in the public domain exactly what procedures the CIA uses when its operatives conduct legal strikes from remotely operated platforms. Because the public tolerates, and even expects, a high degree of secrecy from intelligence agencies, there is far less public oversight of their activities in the war against al Qaeda and other terror groups. Further, the CIA is effectively immune from requests for data and even lawsuits launched under the Freedom of Information Act.

The use of civilian contractors to fly RPAs and conduct lethal strikes without military sanction is in a legal gray area, at best. These operators cannot claim the protections provided by the laws of armed conflict, but because they are in a struggle with an enemy that neither respects nor adheres to those laws, that protection is mostly worthless in the current conflict, and hence it is a distinction without a difference to suggest that contractors are somehow more vulnerable than military operators. Al Qaeda has certainly never professed a desire to follow the Geneva Conventions or to restrict its activities to the battlefield. To the contrary, al Qaeda avoids direct engagements with military forces whenever possible and seeks to conduct mass casualty attacks against civilian targets. Thus, the laws of armed conflict are not an absolute guide to the War on Terror, though many argue that it is imperative that the Western democracies follow them as a means to maintain their legitimacy.

If mercenaries allow the opportunity to engage in lethal activities that would not be countenanced by military forces and remotely driven robots allow attacks on locations that would otherwise be too dangerous to approach, then it is only logical that fully autonomous lethal robots will enable even more extreme activities at the very boundaries of legal and ethical behavior. In some ways, a long-loiter RPA armed with highly accurate munitions represents the greatest precision weapon devised to date. Such a robot can theoretically conduct systematic searches for any potential targets on preapproved lists and then attack them at the first opportunity, with or without the consent of a human operator.

At best, this approach to conflict might push the collateral damage and civilian deaths associated with warfare down to the minimum possible levels. An examination of the gradual progress of lethal RPA strikes shows that they have gradually become more accurate and inflicted fewer unintended casualties over time. As operators become more skilled in the utilization of these weapons and planners become more familiar with their capabilities, it is not surprising that their accurate and effective employment has improved. Fully autonomous robots, or those using some form of distributed artificial intelligence, might prove even more accurate and patient than the human operators currently operating unmanned aerial vehicles and launching airstrikes.

THE HUNT FOR HIGH-VALUE TARGETS

The ability to launch airstrikes with extreme precision from remotely piloted aircraft offered a unique new tool for counterterrorism efforts. In particular, it became possible to hunt individual terror leaders from extreme distances, stalking them at all hours and reacting almost instantly to any confirmed detection. One unsurprising ramification of the new technology was a determination to engage in decapitation operations, on the assumption that killing the key leaders of a terror organization would likely lead to its collapse. At the very least, a significant amount of attrition among organizational leadership within al Qaeda would likely erode the group's ability to plan and carry out large-scale attacks as a follow-up to September 11. As a result, the United States increasingly began to focus upon intelligence collection efforts to identify and target prominent members of al Qaeda. The question of whether these should be considered assassinations remains an open issue—while the killing of ordinary members of al Qaeda probably do not rise to the level of political killing, unmanned aerial vehicle strikes targeting top leadership might. And yet, given that al Qaeda openly declared war upon the United States in 1996 and the United States effectively reciprocated in 2001, there is a counterpoint that al Qaeda members, including the leadership, should consider themselves at war and thus subject to military strikes at all time. Abraham Sofaer pronounced that killing members of al Qaeda should not be considered a form of assassination.[36]

The so-called global War on Terror, known by a host of other acronyms and euphemisms, officially began on September 22, 2001, when Congress passed an Authorization for the Use of Military Force (AUMF). Once President Bush signed it into law, an act he did with a certain determination to rain punishment down upon the enemies of the nation, he had a substantial amount of latitude as to the means, methods, and regions of conflict. To most contemporary observers, the AUMF obviously gave the president the authority to attack al Qaeda wherever the elusive terror network might choose to hide. In particular, this meant that the United States would almost certainly overthrow the ruling party of Afghanistan, where the Taliban government had both instituted harsh sharia law and offered safe haven to Osama bin Laden and his followers. According to RAND analysts Lynn Davis, Michael McNerney, and Michael Greenberg, "Congress has also given the President authority under the Covert Action Statue for covert operations against terrorists presenting an imminent threat. While President George W. Bush's memorandum of notification following the 9/11 attack remains classified, the authorities given the Intelligence Community are reportedly broad and aimed at al Qaeda and any affiliated groups."[37] This would seem to provide at least some legal cover for the CIA to engage in lethal activities. As the capabilities of armed RPAs became apparent, the number of strikes quickly rose, with the effect that "In the first few years of the Obama administration, apparent indifference to collateral damage in Pakistan or Yemen's secret wars only seemed to worsen. Credible reports began emerging that the Agency's Predators were now deliberately targeting rescuers at the scene of earlier drone strikes."[38]

The secular ambition of both al Qaeda and the Islamic State is the formation of a unified Islamic state encompassing the Sunni nations of the world. If such a state could be realized, it would stretch from North Africa to the Pacific and include a population of more than one billion. Its economy would rival that of the other superpowers and thanks to the inclusion of Pakistan, it would possess nuclear weapons. It would also possess well more than 50 percent of the world's known oil reserves. In short, it would be a formidable economic and military power sitting astride a key geographic region—but would it suddenly gain a seat in the most exclusive international organization, the UN Security Council? It is highly unlikely that the existing powers of the world would accede to granting veto power to such a state, and the leaders of al Qaeda are well aware of that fact. The only mechanism for them to become a truly global entity is to pull down the existing system by any means at their disposal.

For their part, the coalition of nations opposing al Qaeda and other terror organizations has no choice but to follow the laws of armed conflict against the non-state groups, no matter how difficult and frustrating that prospect proves. To abandon the legal parameters of warfare would move al Qaeda much closer to its goal of destroying the current world order, as they would be able to point out that the global powers follow their own laws only when convenient. Even if abandoning the laws of war might make the tactical level of engagement against terror organizations a far easier prospect, the long-term political, legal, and ethical ramifications would be devastating. Although the Western coalition is often accused of losing the propaganda fight, particularly in the effort to influence Muslim populations, any proof of overt hypocrisy regarding the laws of war would certainly make such a situation considerably worse.

The 2001 AUMF did not place geographic limits upon the president's ability to conduct operations against al Qaeda, its allies, and its sponsors. While it undoubtedly covered operations against the primary al Qaeda strongholds in Afghanistan, in 2003 it was used in support of President George W. Bush's campaign to invade Iraq. By the time of the invasion, armed variants of Predators had become commonplace, and advancing coalition forces relied heavily upon ISR imagery obtained by RPAs flying overhead. Soon after the fall of Baghdad, Predators and Reapers became the primary observation platforms flying over military convoys, which were increasingly subjected to attacks via improvised explosive devices.

The AUMF preempted any need for President Bush to notify Congress of a deployment of U.S. forces, although he promised to work closely with Congress to set the parameters of the war to come and to make certain that congressional leaders had a say in any key decisions, including any significant expansion of the conflict. Such a partnership not only presented a united front to the world, it also helped to secure future congressional support, including financial backing, should it prove necessary. Previous presidents who wished to take aggressive actions against terror organizations had not always found such a receptive and supportive audience in Congress. Had President Bush called for a declaration of war against any plausible enemy in the aftermath of the September 11 attacks, it is entirely plausible that he would have received it. So why didn't the president seek the ultimate permission to conduct violence in the nation's name?

THE LEGAL ASPECTS OF MILITARY ROBOTS

Military robots do not inherently demand, or deserve, a separate set of legal guidelines. So long as they are under the positive control of a human being, even one extremely distant from the machine itself, they are not fundamentally different from other forms of war machines used by human operators. Thus, the fundamental question of whether or not drones are a legal means of conducting warfare is somewhat of a non-issue. So long as the war that they are being used in is legal and they are used in a legal fashion, they are every bit as legal as other elements of interstate conflict. Regarding the War on Terror, military robots are a legitimate and legal mechanism for engaging in violence as long as their usage conforms with both U.S. domestic law and the international laws of armed conflict.

The AUMF, although somewhat vaguely worded, provides at least some legal cover for the use of military robots. The United States clearly declared an intention to use military force against al Qaeda, related groups, and governments intending to harbor the organization. Although the use of military force is traditionally reserved for military forces, even the flights of RPAs by the CIA and its contractors are at least plausibly legal, so long as they are conducted in a discriminate and open fashion. However, it is not so clear that it is legal for the U.S. government to target American citizens in foreign countries without any significant degree of judicial oversight. If American citizens engage in open military acts against U.S. or allied forces, killing them in the heat of battle is undoubtedly a legal act. If, on the other hand, they are outside of the clearly defined conflict zones and are not engaged in acts of violence, it becomes far more difficult to justify targeting American citizens.

Should Anwar al-Awlaki and Samir Khan have been allowed to claim a certain form of "diplomatic immunity" rendering them safe from all American offensive activity? It is hard to defend individuals who truly are determined to kill their fellow citizens and who are using all of their capabilities to incite acts of violence. On the other hand, placing a wiretap on their cell phones would require the oversight of the Foreign Intelligence Surveillance Court, a branch of the judiciary, while placing them on a list of individuals to be killed at the first opportunity required no such judicial review. Thus, it was effectively easier to kill the two men in question than it was to listen to their phone calls with one another—a situation that calls into question the logic and validity of the kill list, as well as the decision to attack American citizens without making absolutely certain that they should be targeted. To some observers, this approach merely extended an approach that the Obama administration had already adopted, namely, a preference to kill enemies rather than capture them, thus avoiding the need to hold them indefinitely at Guantanamo Bay or other enclosures. As

Jo Becker and Scott Shane framed the situation, "Mr. Obama has avoided the complications of detention by deciding, in effect, to take no prisoners alive."[39] Because the precise mechanisms for placing individuals onto the list, and, for that matter, the question of whether anyone ever comes off the list once added, are considered key national security issues that cannot be publicly discussed, it is impossible to ascertain precisely how much oversight and checks there might be upon the program. The American Civil Liberties Union, which has repeatedly sued (unsuccessfully) to uncover the mechanisms behind the kill list, pointed out that "the U.S. targeted killing program operates without meaningful oversight outside the executive branch, and essential details about the program still remain secret, including what criteria the government uses to put people on CIA and military kill lists as well as how much evidence is required before it does so."[40] David Sanger argues that the program is retained by the intelligence agencies for precisely this reason, as the Department of Defense would be more likely to publicly explain the reasoning behind attacks and to hold the political leadership accountable for mistakes.[41] Chris Woods believes that the Obama administration, while eschewing judicial intervention, at least provided some information to congressional leadership, claiming "While Obama, like Bush before him, had refused to make either JSOC or the CIA publicly accountable for their targeted killings, he did appear to place some faith in secret congressional oversight."[42]

From an international-law standpoint, the use of military robotics falls under the same category as the use of a manned vehicle for an attack. If a conflict is sanctioned by the UN Security Council, is carried out in the face of imminent attack, or is a reasonable form of self-defense for a nation, the means of conducting the military attack is irrelevant. In the case of the wars in Afghanistan and Iraq, there is some degree of international agreement that both locations constitute an armed conflict zone, one in which combatants should expect to be subjected to attack at any time and, hence, the use of surprise attacks via unmanned platforms is almost certainly legal (though perhaps not in every sense). The use of armed drones outside of Afghanistan and Iraq, on the other hand, becomes more problematic, particularly when they are being used to attack terror organizations and individuals who are only marginally connected to al Qaeda. As John Sullins points out, "Typically, the missions that are flown by intelligence agencies like the CIA are flown over territory that is not part of the overall conflict. The 'War on Terror' can spill out into shadowy government operators engaging an ill-defined set of enemy combatants anywhere on the globe that they happen to be."[43] And, of course, the rest of the world is looking to the United States to establish the precedents for the use of unmanned vehicles, with an eye toward their own capabilities and interests. Providing the legal precedent for intelligence agencies to engage in large-scale lethal operations is likely to breed more chaos in the immediate future.[44]

The War on Terror amply demonstrates, regardless of the weapons platform in question, that the law almost always needs time to catch up with new forms of military action. The international laws of armed conflict tend to be an after-the-fact reaction to the most recent (and concluded) conflict rather than an anticipatory system that can be adapted to every potential new scenario. When Francis Lieber devised his code of conduct or the participants in Geneva agreed upon the conventions of modern warfare, no one expected to see the most powerful state on earth engaged in a fight with a non-state actor. In part, this is because of the obvious mismatch of the two combatants—but in part, this is because it did not seem possible that a non-state actor would deliberately engage in such a conflict and be able to continue the struggle for more than a brief period of fighting. By surviving, and even striking back upon occasion, al Qaeda and similar groups not only inspire the next generation of terrorists, they also demonstrate the very difficult prospect of destroying an ideology or making war upon a form of behavior.

Many legal scholars have argued that the use of military robots outside of conflict zones is illegal. For example, Mary Ellen O'Connell posited, "There was no armed conflict on the territory of Pakistan because there was no intense armed fighting between organized armed groups. International law does not recognize the right to kill with battlefield weapons outside an actual armed conflict. The so-called 'global war on terror' is not an armed conflict."[45] Her point, of course, is that the decision to strike into Pakistan, whether with manned or unmanned platforms, was inherently illegal, not that the use of military robots is in and of itself against the laws of war. However, the prevalence and utility of military robots seem to be drawing successive administrations into the belief that military action is the first resort when dealing with terrorists, regardless of location, an idea that would likely have been unthinkable without the new technology. As Human Rights Watch put the matter,

The notion that the entire world is automatically by extension a battleground in which the laws of war are applicable is contrary to international law. How does the administration define the "global battlefield" . . . ? Does it view the battlefield as global in a literal sense, allowing lethal force to be used, in accordance with the laws of war, against a suspected terrorist in an apartment in Paris, a shopping mall in London, or a bus station in Iowa City?[46]

Of course, the legal aspects of international conflict are not the only governing force regarding the future use of military robotics. Just because an action is technically legal does not inherently make it an acceptable act—it must also be morally justified. This is especially true with regard to acts of war, as the death of human beings cannot be undone and should not be encouraged without both legal and moral considerations. While military robotics tend to push the boundaries of legal warfare, they can usually be incorporated into existing systems for most applications. The same is not necessarily true regarding the moral limits of international warfare.

CHAPTER 6

Morality for Machines?

I shall not consider anyone of you a fellow-soldier of mine, no matter how terrible he is reputed to be to the foe, who is not able to use clean hands against the enemy. For bravery cannot be victorious unless it be arrayed along with justice.

—Flavius Belisarius, c. 533

By definition, robots are not influenced by the passion of violent emotions, although the humans that send them forth to engage in violence are certainly not immune from such emotions. Robots are no more susceptible to passion than any other machine, and there is no compelling reason that emotions will, or even should, be incorporated into robotic systems. Robots are theoretically as vulnerable to the effects of chance and probability as any human engaging in warfare is, although the computer systems that substitute for brains in robots will probably have a much better likelihood of correctly predicting the probable outcomes for certain actions and their reaction times can be far quicker than that of humans. As for the role of reason in the conduct of warfare in pursuit of political gains, robots are certainly governed by a form of logic, one that is programmed into their very core and thus directly drives their patterns of behavior. However, any such logic system, if it is too rigid, creates the possibility of a predictable behavior within the systems in question and thus introduces a vulnerability that might be exploited by an enemy. Thus, the more autonomous a machine, the more unpredictable its behavior, although such unpredictability comes at a cost of the remote operator's ability to specifically order or prohibit certain actions.

Modern warfare is largely bounded by a system of international agreements that set limits upon the types of acceptable behaviors in wartime. Of course, the system is not perfect, even though it has gradually evolved over nearly four centuries, and its practitioners cannot be counted upon to follow the rules on every occasion and in every situation, even if the rules were so clear-cut as to make such adherence possible. In this regard, military robots might actually have a significant advantage over their human counterparts, one that could serve to mitigate some of the worst aspects of modern warfare. Given the vital nature of programming in creating an acceptable system of behaviors within a robot, if all of the laws of warfare could be programmed into military robots, it might be possible to essentially prohibit violations of the laws of war, an outcome that would surely diminish the worst aspects of warfare.

Codifications of acceptable behavior in wartime go back for millennia, although for virtually any system of laws, customs, or expectations, there have been exceptions made. Most of the time, those exceptions were applied to "barbarians," essentially announcing that the rules applied only if every party to the conflict agreed in advance to the limits of the violence. There are obvious benefits for creating such systems of limits—even when two or more societies feel the need to engage in violence and killing in pursuit of their political objectives, that does not mean that the wanton slaughter of the entire enemy population is an acceptable, or even desirable, outcome. In particular, wars of conquest have rarely been limited to the territory and natural resources of an area—they have normally also included the inhabitants of the region, who are expected to gradually assimilate into the new system, perhaps grudgingly, but inevitably, over a space of generations. The largest empires of human history have largely been created upon this concept—none of them managed, or even significantly tried, to maintain some form of racial purity across the population, although different levels of "citizenship" have certainly been extended to conquered populations.

This large body of international thought regarding the conduct of warfare only makes sense if wars are conducted between humans. As more robots are introduced to the battlefield, the ethics become more problematic. Morality is a uniquely human concept. Rarely can morals be simplified to a set of ironclad rules, though moral behavior requires individuals to weigh their potential actions against their moral code and decide how to behave in ambiguous situations. Wartime, in particular, creates an enormous set of morally hazardous options, particularly as they apply to the use of lethal force.

THE PRINCIPLES OF ETHICAL WARFARE

Ultimately, the ethics of armed conflict, which govern the morally permissible acts that might be committed once a state of war exists, come down to a handful of key principles.[1] These principles include proportionality, discrimination, and necessity. Proportionality suggests that a military action taken in response to an enemy act should be similar in size, scope, and violence. Thus, a small border incursion should not be met with a full-scale nuclear attack as a retaliatory action—such a response would be disproportionate in the extreme. However, the notion of proportionality in war does not require that the victim respond in precisely the same fashion as the original aggressor—not only might such a response be impossible, depending upon the means at the responder's disposal, but it might also be ill advised depending upon the preferred modes of conflict adopted by the enemy. Essentially, if proportionality required a response to be a mirror-image of the original act, aggressors would have the benefit of setting the limits of a conflict to their most advantageous position. A dominant sea power might then force a land power to fight only on the oceans and thus gain an enormous advantage, despite being the originator of the conflict.[2]

Proponents of the creation, production, and utilization of military robots argue that robots, unlike their human counterparts, can be expected to completely follow any moral and ethical guidelines without reservation.[3] Naturally, such behavior will require that predefined ethical constraints be codified and translated into a machine language, which is not a simple task. In order to create such a system, military robots will either require a core set of instructions that cannot be violated under any circumstances or need a powerful artificial intelligence to interpret moral guidelines that might occasionally conflict.

The task of translating moral and ethical codes is one that human organizations, in particular religions, have struggled to perform for millennia. Most of the world's largest religions have expanded well beyond the geographic region of their origin, and as new converts join the faith, they need to be instructed in its requirements for all adherents. Expecting would-be converts to learn a new language as a means of acquiring religious instruction is not a particularly effective strategy for propagating a religion. For most (though not all) faiths, the faster method has been to translate the holy dogma into target languages and to send members of the faith to proselytize in that same language. Given the wide variety of languages in the world, it is unsurprising that a certain degree of alteration might occur in the act of translation. Perhaps the new language does not contain the necessary terms for an exact translation, leaving the translator to select between the literal and the essential to communicate the religion's concepts. Most faiths account for this by essentially having an "official" or sanctioned version of the faith's dogma and admittedly imperfect versions in other languages. For example, within the Roman Catholic tradition of Christianity, the Latin Vulgate Bible was the standard for centuries, even though the first Biblical texts were written in Hebrew, Greek, and Aramaic. Starting in the 17th century, it was gradually supplanted by the King James Version in English-speaking countries. Within Islam, only the original wording of the Q'uran, in Arabic, is considered authentic—and converts who wish to more deeply understand their faith must learn Arabic to read this official version. However, the Q'uran has been translated into hundreds of languages, at least opening the possibility of sharing the faith with potential converts, who might then be induced to study it in a deeper sense in the original Arabic.

One of the most oft-cited religious sources of morality and ethics appears in the Old Testament of the Christian Bible and the Torah of Judaism. The Ten Commandments, presented by the prophet and religious leader Moses to the Israelites, are written in the Book of Exodus, Chapter 20, verses 1 to 13. Originally provided in Hebrew, they were translated into Greek, then Latin, then English for the King James Version of the Bible, first published in 1611. Comparing the wording of the commandments through the various iterations illustrated the problem of translating moral precepts, without accounting for how to explain them in a fashion a machine can understand and follow. The first commandment is usually referred to as "Thou shalt not kill." Yet, the Bible is replete with killing, often upon the order of God, particularly in the Old Testament, which might suggest either a flaw in the early adherents' faith or an error in translation. Further complicating the issue, an absolute prohibition on killing would theoretically act as a ban upon survival for humans, as it would be next to impossible to survive in the world without at least occasionally killing food sources, even if only through accidental means. Modern scholars have long argued that a more accurate phrasing, based upon earlier written records, reads "Thou shalt not murder." This more accurate translation makes substantially more sense. Not only is it impossible to murder anything other than another human, not all forms of human killing constitute murder. Thus, this commandment did not prohibit all killing; it only served to clarify when killing is or is not an acceptable act. The more modern translations may have served the potentially useful purpose of encouraging the faithful not to kill any other members of society. They may have also assisted in the pacification of native populations, as forced conversions would have essentially demanded that the new converts stop any violent efforts against the Christian invaders. Regardless of any potential social benefit, the prohibition of killing conflicts with the practice of warfare—and yet ostensibly Christian nations have engaged in warfare as much as those of other faiths.[4]

To instruct machines that they are not to kill human beings would be a fairly simple set of instructions, although it would certainly limit the martial effectiveness of most military platforms. Machines, unlike their human counterparts, do not question their instructions or wonder why there might be an absolute prohibition on killing. Accidents would still occur, of course, due to the unexpected circumstances that always arise in wartime, but it would be easily provable that the actions of the robot were inadvertent—in attempting to avoid killing, it might accidentally cause a different death. Yet, the pipe dream of conducting warfare without killing is completely unrealistic, particularly if all parties to a conflict do not agree to the same rules. Just as some belligerents in modern conflicts adhere to the laws of war and some do not, the same is true of moral codes. Those choosing to ignore the requirements of legality or morality might even garner a short-term advantage over the adherents, although the public opprobrium their actions earn might offset those short-term gains.

Chinese military theorist Sun Tzu, writing in the sixth century BCE, offered his thoughts on how military commanders could successfully undertake warfare and achieve the desired political ends at minimal cost to their own forces, in his seminal work *The Art of War*. His ideas were primarily offered as a series of single-statement dicta, making them easily digestible pieces of advice that have retained a certain timeless quality. Admonishments such as "For to win one hundred victories in one hundred battles is not the acme of skill. To subdue the enemy without fighting is the acme of skill" are difficult to debate in theory, although Sun Tzu offers far less practical guidance in how to achieve such an end.[5] Yes, it would be an impressive achievement to subdue an enemy without engaging in a bit of violence, and, in some cases, merely presenting an unwinnable situation to the enemy might be enough to provoke a surrender, given that the alternative might be a certain death. Perhaps his most well-known statement regarding the art of war is "Therefore I say: 'Know the enemy and know yourself; in a hundred battles you will never be in peril.'"[6] On other occasions, Sun Tzu repeated the admonition that a general's self-understanding was far more important than any knowledge that might be gained about the enemy, essentially calling for a realistic assessment of one's own forces and their capabilities. These ancient aphorisms might still have value for a modern or future conflict conducted primarily by autonomous machines.

In much the same way that Sun Tzu recommended for human combatants, it might be most effective to maneuver in warfare in such a fashion as to demonstrate the futility of resistance before any actual violent acts are conducted. Autonomous machines might offer threats of violence without being forced to carry them out, and the mere demonstration of their capabilities may be enough to cow the enemy before any death or destruction become necessary. Yet, suspecting that an enemy might be able to inflict casualties upon your own forces or upon your civilian populace has never proven a particularly effective mechanism to compel a surrender. Human societies do not simply give up on a cause that they feel compelled to defend by force on the basis of fearing they might lose a conflict. Weaker societies often join mutual protection alliances in part because it offers a way to counter the potential predatory behavior of stronger neighbors. Even in situations when such alliances have not been possible, weaker states have derived a certain amount of protection by working to make certain that an attack will not be worth the costs involved, in part because it might open the stronger opponent to attack by another large rival. If societies were so fragile that they might give in to the pressure of an inevitable defeat, the entire history of warfare since 1945 would have been entirely different. How could a non-nuclear power, and one with no alliances of note that might offset that disadvantage, possibly stand up to nuclear-armed powers? Yet, in dozens of conflicts since the advent of nuclear weapons, the stronger, nuclear-capable nations have lost when engaging in wars against much weaker and less technologically developed nations. How might Sun Tzu explain the Soviet Union's failure to defeat the relatively weak and disorganized resistance offered by mujahideen in Afghanistan or the struggles of the United States to defeat first the Viet Cong and then the North Vietnamese, in Southeast Asia? Was it because neither the Soviets nor the Americans knew themselves well enough to conduct their wars? And further, neither felt the need to spend much time, if any, obtaining a working knowledge of the enemy they faced; the assumption of inherent superiority may have carried with it a military hubris that could not be offset by any level of technological development. Is it rational to believe that the creation, procurement, and fielding of military robotics will provide a greater advantage than that offered by nuclear weapons? Or is it more likely that these autonomous machines will only alter the tactical conduct of warfare, and, while they might reduce the bloodshed for the nation that employs them, they will not guarantee a victory in conflict any more than any earlier weapon has ever done?

During the Thirty Years' War, European nations warred back and forth across the continent in a conflict that eventually drew in nearly every Continental power, large and small. At issue was the religious schism between Catholic and Protestant Christians, although political, cultural, geographic, and economic factors all played a substantial role in the origins and the extensions of the conflict. During the three decades of warfare, very few limits governed the acceptable behavior of the warring parties. The religious aspect of the warfare certainly contributed to the problem—each of the warring sides felt it operated under the mandate of the supreme deity and that any punishments visited upon the enemy were perfectly justified ways of punishing heretics. Although taking prisoners and holding them for ransom or exchange was a fairly common occurrence during the war, so was the execution of enemy troops seeking quarter. Observing the carnage, Dutch jurist Hugo de Grotius, who had experience as a prisoner, wrote *De Jure Bellis ac Pacis*. This three-volume treatise examined the question of when a nation was entitled to undertake war and how wars must be fought. In the process, the treatise established the concept of Just War, a theory that states do not have an inherent right to engage in warfare without justifiable circumstances and that there are limits to what kinds of activities could be practiced even if a war was allowed. Grotius's work established the concept that the rules of warfare applied to all belligerents, regardless of whether or not they considered themselves bound by the rules. It also clarified that wars are fought by governments through the use of military forces, and, as such, the decisions made in warfare belong to the nations at war, not to the individuals involved in the fighting. Thus, if prisoners were taken in a conflict, for example, they belonged to the capturing state and could not be ransomed by the units that effected their capture. Likewise, any property confiscated in war belonged to the state, not to the individuals taking it.

Grotius's ideas quickly gathered steam, and a new understanding of the acceptable practices of warfare swept through Europe, and, by extension, to other regions of the world where European armies clashed. However, as had been the case for many previous codifications of warfare, the presumption was that the European rules of warfare only applied to conflicts between "civilized" nations, a delineation that was presumed to extend only to other Europeans. Thus, no limits might be recognized for a fight between a "civilized" power and a "savage" population, meaning the wars of colonial conquest might be conducted with unfettered violence and no fear of retribution. While his concepts had a civilizing effect upon European warfare and established norms that continue to govern modern warfare, that civilizing effect was itself quite limited then as the major European powers embarked upon a massive wave of colonial expansion, most of it accompanied by unprecedented levels of violence.

The ideas of Grotius were not the only source of attempted expectations of ethical behavior in warfare. Other scholars of subsequent centuries contributed to the theories of how warfare should and should not be conducted. The most prominent and influential included Montesquieu, Vattel, and Rousseau. The complete examination of each of their contributions is beyond the scope of this work, but suffice it to say that the general trend from the mid-17th until the mid-20th centuries was to increasingly limit the legitimate mechanisms of conducting warfare, often in response to technological and doctrinal developments that pushed beyond the boundaries of earlier limits. At the same time, an increasingly explicit body of regulations began to lay out the expectations of belligerents engaged in warfare, rather than simply attempting to prohibit extremely deleterious behavior. Of course, these legal arguments continued to apply almost exclusively to the European states and included only a limited number of non-European nations, mostly those of former European colonies that had declared independence. Not until the 20th century did those limits get extended around the globe through a series of international agreements, such as the conventions signed at The Hague (1899 and 1907) and Geneva (1929 and 1949).

The great Prussian military theorist Carl von Clausewitz posited that warfare is comprised of three basic elements. The first is the passion of violent emotion; essentially, there needs to be an underlying cause for the conflict in the first place, one that means enough to individuals that they would choose violence as a means of responding to the situation. The second is the interplay of chance and probability, which is often reduced to the notions of luck and circumstance, demonstrating that no amount of planning or preparation can guarantee a specific outcome. The third element is political calculations driven by reason, illustrating that warfare is not simply an exercise in anarchy but rather an act characterized by violence that is bounded by the rational decision making of its practitioners. Clausewitz's masterwork, *On War*, is widely regarded as one of the most important treatises on military theory and has been cited as the best explanation of the fundamental nature of war that has ever been created. An imperfect understanding of Clausewitz's ideas, largely by reducing them to a series of dictums, has been a key element of professional military activity for the nearly two centuries since his work first appeared and its translation into dozens of languages has expanded its influence to every corner of the globe. Yet, many of the concepts of Clausewitz that have proven so useful to understanding human conflicts may not be applicable to warfare conducted by autonomous machines.[7]

The Hague and Geneva Conventions are far more explicit than the general pronouncements of individual writers. Both prohibited the use of certain weapons, established the notions of who was considered a legal combatant (and hence a legitimate target in warfare), and created expectations of how to treat captured enemies. The Hague Conventions, signed prior to World War I, had a provision that they would go into effect only if all parties of a conflict were signatories to the agreement, demonstrating the authors' belief that a general conflagration like World War I would be impossible in the modern era. Because Serbia and Montenegro were both parties to the conflict but not to the Convention, the requirements of the treaty technically did not apply to that war. Thankfully, the requirements of the agreement offered such obvious advantages to all belligerents and did so at a relatively minor cost in resources and effort that most of the parties to the conflict substantially lived up to the agreements. However, the potential for a collapse in such an informal agreement demonstrated the need to for a new system of limits upon warfare and led to the Geneva Convention of 1929. That meeting led to a series of international protocols that, among other things, formally banned the deliberate targeting of civilian populations (although some degree of collateral damage is inevitable in war) and required all prisoners of war (POWs) to be treated in a humane fashion. In exchange for protection from the worst horrors of warfare, civilians are formally banned from engaging in armed conflict, even during a war, and should they do so, they become legitimate targets for the enemy. If they should inhabit territory that becomes occupied by the enemy, they are expected not to engage in acts of sabotage or guerrilla attacks, lest they forfeit the legal protections afforded by the agreement.

Likewise, POWs can claim legal prisoner status only if they conform to the rules of warfare, including bearing arms openly, wearing uniforms, and following an identifiable chain of command. Failure to follow these regulations renders one an illegal combatant who is no longer entitled to protection from harm, shelter, and food, and, if needed, medical care.

PROPORTIONALITY

Given the long history of human conflict, there are an almost inexhaustible number of examples that might be used to illustrate proportional and disproportional attacks. The vast majority of military operations, particularly once an established state of warfare exists, can be justified as proportional and therefore unexceptional. The opening stages of a conflict certainly offer the opportunity for a disproportionate response, and it is a mark of restraint when a nation limits its initial actions to proportional ones. This restraint may serve as a useful function of preventing a cycle of escalation, or it may be due to an inherent desire to fight in a moral fashion, but the net effect is the same.[8]

When North Vietnamese torpedo boats attacked the USS *Maddox* in the Gulf of Tonkin on August 2, 1964, President Lyndon B. Johnson clearly considered it a hostile act. Yet, he did not seek an outright declaration of war upon the Democratic Republic of Vietnam; he instead requested formal permission from Congress to react in a proportionate fashion. On August 7, Congress passed the Tonkin Gulf Resolution, which specifically authorized the president to "take all necessary steps, including the use of armed force, to assist any member or protocol state of the Southeast Asia Collective Defense Treaty requesting assistance in defense of its freedom."[9] President Johnson, operating with military advice, opted to launch Operation Pierce Arrow on August 5. This series of aerial attacks launched strikes against North Vietnamese torpedo boat bases and petroleum depots. In March 1965, Operation Rolling Thunder expanded the targets, including the destruction of air defense radar sites and their related surface-to-air missile batteries and air defense artillery. It commenced a 44-month campaign characterized by gradual escalation and an unwillingness to risk provoking Soviet or Chinese intervention. Of course, a first-rate airpower such as the United States could have immediately engaged in a disproportionate campaign of attacks, up to and including nuclear strikes against major population centers or attacks to compromise the Red River dams. Such attacks against North Vietnamese infrastructure had the potential to devastate the countryside by unleashing widespread flooding, probably inducing a humanitarian crisis with an entire growing season's crops washed away. Despite the fact that such an attack comprising of airstrikes would likely have killed only a handful of Vietnamese citizens, the secondary effects of such a campaign would have been horrendous. At the same time, they likely would have ended the war in short order.

In the aftermath of Operation Desert Storm, the United States and its coalition allies established a pair of no-fly zones in northern and southern Iraq, largely to prevent the Iraqi government from carrying out aerial attacks against segments of its own population. Operations Provide Comfort, Northern Watch, and Southern Watch involved the commitment of dozens of military aircraft to each zone and the deployment of thousands of personnel to support them. They lasted from 1991 to 2003, and over that time the Iraqi military occasionally engaged in provocative, hostile acts, such as illuminating coalition aircraft with air defense radars. Rather than resuming full-scale hostilities, the coalition forces tasked with carrying out Northern and Southern Watch restricted themselves to proportionate responses, including carefully targeted precision strikes against the offending radar sites.

Disproportionate attacks, while far less common, tend to stand out by their very nature, particularly because they often include enormous loss of noncombatant lives. One of the lesser-known but still prominent examples of a disproportionate response to provocation came in 1898. At that time, the last remnants of the Spanish Empire were crumbling—most notably in a massive Cuban insurgency that threatened to drive the Spanish out of the Caribbean. American spectators, whipped into a frenzy by jingoistic newspaper accounts of Spanish atrocities in Cuba, created a fertile environment for American military intervention. President William McKinley dispatched the USS *Maine* to Havana Harbor as a means of protecting U.S. citizens in Cuba and maintaining observation of the ongoing conflict. On February 15, 1898, the *Maine* exploded, due to then-unknown causes.[10] A rushed investigation concluded that Spanish agents, for unknown reasons, had emplaced an explosive device on the hull of the *Maine*. When it detonated, it triggered a series of secondary explosions, sinking the ship and killing 266 of its crew.

The American public reacted with seething rage, demanding that the U.S. military be unleashed to punish the Spanish for their treachery. Swept up in the tide, McKinley requested a declaration of war, which he received on April 25. To demonstrate the high-minded ideals behind the declaration, the wording of the document included a specific amendment guaranteeing that the United States had no territorial ambitions in Cuba, although it made no such guarantees about any other Spanish possessions. In the ensuing naval battles, the utter technological superiority of American battleships was displayed without leaving any doubt for the spectators. At Santiago, Admiral Pascual de Cervera's lost his entire fleet, losing 5 ships and 323 sailors while inflicting almost no damage to its U.S. opponents. On May 1, Commodore George Dewey sailed into Manila Bay and attacked another Spanish fleet at anchor. Seven hours of U.S. shelling destroyed that fleet as well, inflicting a loss of 8 ships and 77 personnel. During the action, Dewey's force lost one sailor, not due to enemy fire but due to sunstroke. In short, the naval fights were more a slaughter than a battle. American negotiators all but imposed the Treaty of Paris upon Spain, freeing Cuba but obtaining Puerto Rico and the Philippine Islands as American protectorates. The level of destruction unleashed by American warships in response to a dubious provocation was entirely outside the bounds of proportionality, and many commentators have noted that the United States, having officially "closed" its internal frontier, had begun to engage in imperialism like its European competitors. If true, it is unsurprising that a nation with imperialist designs would utilize the slightest provocation as an excuse to engage in a war it expected to win at minimal cost.

In 1937, Japanese troops attempting to conquer China laid siege to Nanking, a city of more than one million inhabitants serving as the temporary national capital. When the defenders of the city refused the initial Japanese demand to surrender, the Imperial Japanese Army, commanded by General Iwane Matsui, pressed a vicious attack upon the city. The Japanese made no effort to show restraint in the assault, and, once the city's defending forces had been annihilated, Matsui allowed his troops to sack the city. The result was an orgy of violence, with tens of thousands of Chinese civilians subjected to countless atrocities, including murder, rape, and torture. International observers could do nothing but watch as the Japanese Army systematically looted and destroyed the city with a level of violence not seen in China for over 700 years. When the pillaging finally ended, most historians studying the "Rape of Nanking" placed the civilian death toll in excess of 100,000. The Japanese offered little by way of justification for their behavior, beyond the vague suggestion that the city should have surrendered when offered the chance and that it had brought about its own punishment and desolation for daring to resist the Japanese advance. Again, any sense of proportionality evaporated in the aftermath of the city's collapse—even if the Japanese had chosen to destroy the city's defenses, there was no justification for the events in the aftermath of the battle.

Military robotics offer an interesting dynamic regarding proportionality. At one end of the spectrum, military robots might offer the greatest precision weapons devised to date and thus enable each and every military attack to be of precisely the proportion desired by operational planners.[11] However, unscrupulous (or unskillful) planners might wind up at the opposite end of the spectrum, leading to robots being used to engage in completely disproportionate responses, up to and including genocidal actions. Robots do not inherently determine right from wrong—they follow their programming—and they do not tire of the same activity repeated an unlimited number of times, including killing. By definition, they will not become horrified by their actions or ponder the morality of their actions—they will simply do as they have been told to to the limits of their power supplies.[12]

DISCRIMINATION

Discrimination requires that a belligerent be able to distinguish between legitimate and illegitimate targets.[13] Thus, if an enemy force is operating in the vicinity of a civilian population center, for example, it would not be militarily justifiable to use a massive aerial attack to destroy the entire population center, even if that attack would also destroy the military force in the region. Attacks that indiscriminately destroy without regard to the casualties are considered unjust and must be avoided by a belligerent who wishes to fight under the rule of law. This is arguably the most difficult principle of Just War to accurately undertake, in part because of the types of conflict that constitute most modern wars. Guerrilla forces, in particular, derive much of their sustenance from living within the civilian population and emerging on a periodic basis to launch attacks. By hiding within the civilian population, they seek to render themselves immune from an overwhelming conventional attack that might be launched by a presumably much more powerful and technologically advanced foe. However, it is the guerrillas who are inherently placing the civilians in the direct line of fire and bear much of the responsibility for the resulting consequences.

Like proportionality, discrimination tends to be more noteworthy when it fails to occur than when it occurs. Discrimination is usually a more tactical decision, whereas proportionality usually depends, at least in part, upon the strategic decisions of a nation at war.[14] At its most basic level, discrimination requires a belligerent to take all reasonable measures to avoid inflicting civilian casualties and to refrain from deliberate attacks upon civilian populations. Of course, warfare is not conducted with a perfect ability to see and strike targets and some degree of collateral damage, including civilian deaths, is almost guaranteed. Nevertheless, modern combatants attempting to fight in a moral fashion must take steps to discriminate between targets wherever possible. Once again, a pair of examples illustrating successful and unsuccessful discrimination should serve to demonstrate the major point of this aspect of morality.

Modern technology has, in many ways, offered a much greater capability to discriminate between targets. One prominent example is the development and employment of precision-guided munitions (PGM). These weapons can be employed in a wide variety of fashions, and they may be guided using satellite telemetry, laser designators, control wires, or other means. Aerial bombs using PGM technology came to the public's attention during Operation Desert Storm. Military briefings to the media during the 42-day aerial campaign often included footage of PGMs striking very small targets. One well-received example showed a bomb dropping through the ventilation shaft of an Iraqi military headquarters. Its explosion inside the building was largely contained, making it a devastating strike for anyone inside the structure but completely harmless to the civilian buildings on either side of the target. Although PGMs impressed and inspired the public, they constituted only 7 percent of the total number of aerial munitions used in the Gulf War. However, only 12 years later, 70 percent of the bombs dropped in Operation Iraqi Freedom were PGMs.[15] One side effect of such accuracy is that both the public and the enemy come to expect an almost impossible level of precision, and when something goes wrong the recriminations can be brutal. Thus, when an American bomb struck the Chinese embassy compound in 1999, killing three Chinese journalists, most observers assumed it had been an intentional strike rather than the result of a technological malfunction or an intelligence failure.[16]

In 2003, the United States led a multinational coalition to remove Iraqi president Saddam Hussein from power. The initial advance upon Baghdad moved quickly, although Hussein evaded capture until December 13. During the nine-month manhunt, Hussein managed to orchestrate a massive insurgency against coalition troops occupying Baghdad and other key Iraqi locations. Even after Hussein was captured, the insurgency continued to expand and threatened to embroil the entire country in a civil war based almost entirely upon sectarian lines. One major hotspot of insurgent activity was the city of Fallujah, located 40 miles west of Baghdad. In November 2004, coalition troops led by 6,500 U.S. Marines approached Fallujah with the intention of capturing it and removing its status as an insurgent stronghold.

General John Sattler initially maneuvered his forces into blocking positions, essentially surrounding the city rather than simply driving forward in a potentially bloody offensive. Having secured the region around the city, he then called upon the civilians of Fallujah to evacuate, promising them safe passage through coalition lines.[17] Essentially, Sattler was allowing the population of Fallujah to self-discriminate; those who refused to leave were not immediately classified as hostile or treated as targets but they were certainly viewed with a great deal of suspicion. Once the deadline for evacuation had passed, Sattler ordered his troops forward—yet, rather than relying upon artillery and airpower to essentially blast their way into and through the city, his forces engaged in a methodical advance, engaging the enemy in a street-to-street and house-to-house battle for the city. This allowed the advancing forces to protect civilians who had not evacuated the city. The campaign to retake Fallujah cost 95 U.S. troops their lives, but it also prevented what could have been a humanitarian disaster while also refusing to cede a sanctuary city to Iraqi dissidents.

Much like disproportionate military attacks, nondiscriminatory attacks tend to stand out. During World War II, the United States and Great Britain marshaled their airpower to conduct the Combined Bomber Offensive (CBO). This entailed committing tens of thousands of aircraft and millions of troops (including aircrews, maintenance personnel, and support forces) to a sustained bombing campaign over Europe. Its ostensible purpose was to erode the Germans' ability to continue the war by destroying the German industrial capacity. The U.S. Army Air Forces (USAAF), commanded by General Henry "Hap" Arnold, believed that it could conduct high-altitude precision strikes using heavy bomber aircraft equipped with the Norden bombsight. The aiming device had proven relatively accurate in testing, allowing air crews to drop their bomb loads at an average distance of 400 feet from the targeted point when flying at an altitude of 15,000 feet. Of course, wartime conditions were rightfully expected to erode accuracy, but the USAAF planners still believed in the efficacy of pinpoint targeting. The Army War Plans Division created an extremely detailed air war campaign plan, AWPD-1, that envisioned fleets of bomber aircraft devastating German industry while flying high enough to be mostly immune from antiaircraft fire or aerial interception. It called for the production of more than 2,000 B-29 bombers, with a further 3,700 even heavier bombers to be created before the end of the war. The planners estimated that less than 9,000 fighter aircraft would be needed for the entire war effort.[18]

As Helmuth von Motlke famously opined, "No plan survives contact with the enemy." AWPD-1 proved no exception to this dictum—not only did its assumptions about bombing accuracy prove to be hopelessly optimistic, the number of Allied aircraft lost to enemy interceptors and ground fire threatened to end the CBO in short order. On August 17, 1943, more than 300 USAAF B-17s launched a massive strike against a German ball bearing plant at Schweinfurt and an aircraft assembly plant at Regensburg. Poor timing and bad weather, coupled with heavier-than-expected fighter cover and ground-based defenses, resulted in 60 B-17s shot down, with a further 60 damaged beyond repair, all for relatively minor damage to the targets. Even with the enormous U.S. industrial capacity, these types of losses would have destroyed the USAAF in a matter of a few weeks. Realizing the fundamental flaws in their assumptions, Allied commanders called a halt to the bombing campaign to study its progress. The British, who had never accepted the notion of high-altitude precision bombing raids conducted during the daytime, recommended that the USAAF shift their focus from pinpoint attacks against critical industrial nodes to nighttime attacks against entire German cities. These raids were characterized not as attacks against the civilian population, which would be both indiscriminate (and hence immoral) and illegal, but attempts to "dehouse" the German workers. The wordplay fooled no one—the attacks served notice that the Allies had effectively abandoned the principle of discrimination in favor of a strategy designed to kill and coerce civilians to stay away from war industries.

Then lieutenant colonel Robert McNamara, who later served as secretary of defense under Presidents John F. Kennedy and Lyndon B. Johnson, remarked after the fact that U.S. aviators and especially their commanders might have been considered war criminals for the indiscriminate bombing campaigns conducted during the war. In the skies over Japan, General Curtis LeMay not only embraced a strategy of area bombing, he enhanced its effectiveness through the use of incendiaries designed to ignite hundreds of fires in Japanese buildings largely composed of wood and paper. By abandoning discrimination, the Allies undoubtedly brought a quicker end to the war—but they did so at a tremendous moral cost.

One of the more well-known incidents of indiscriminate violence in the modern era occurred on March 16, 1968, during the height of the Vietnam War. U.S. Army helicopter pilots reported antiaircraft fire from the vicinity of My Lai, a small coastal village in South Vietnam. U.S. Army second lieutenant William Calley and the platoon under his command moved into the village on foot and demanded that village leaders surrender the Viet Cong guerrillas Calley assumed must be hiding in the area. The villagers denied any knowledge of the attacks and professed their loyalty to the South Vietnamese government, but Calley refused to believe their story. What happened next horrified Army investigators later sent to the area. More than 300 civilians were killed in an orgy of violence. Despite the presence of American eyewitnesses who reported the incident, the My Lai massacre went essentially ignored for almost 18 months. Calley was eventually prosecuted by a court-martial for the premeditated murder of more than 100 civilians, for which he was eventually sentenced to life imprisonment. President Nixon immediately commuted the sentence to house arrest, pending appeal, and after three-and-a-half years of house arrest Calley's sentence was commuted to time served.

The indiscriminate bombings of World War II and the killings in the My Lai massacre illustrate one truly terrible aspect of indiscriminate killing in warfare: the perpetrators are rarely prosecuted, despite the ample evidence of their crimes. Indiscriminate killing tends to be classified as a regrettable incident or a terrible necessity to expedite the end of a war, not a deliberate act. Sometimes, the nation whose citizens become victims of indiscriminate attacks is blamed, either for starting the conflict or for continuing it despite an inability to protect its own people. This attitude merely confirms that military powers often ignore the needs of morality in favor of military expedience. This observation is not a new concept. Thucydides illustrated the same principle more than 2,000 years ago in his *History of the Peloponnesian War*. When describing a conversation between representatives of Athens and the island of Melos, now known as the Melian Dialogue, Thucydides ascribes to the Athenians the realist statement, "You know, as we do that right, as the world goes, is only in question between equals in power, while the strong do what they can and the weak suffer what they must."[19] While undoubtedly true to a certain extent, it is hardly a relief to know that the challenges of moral behavior in warfare have been a constant problem for millennia.

As is the case with proportionality, military robotics again offer mixed promises regarding their ability to discriminate on the modern battlefield. Thus far, RPAs have been used in a number of questionable fashions, even with a human being at the controls. In particular, the use of signature strikes, in which targets are attacked for engaging in a certain pattern of behavior regardless of the lack of a positive identification, proves extremely troubling when considering whether the War on Terror is being fought in a just manner. In the view of Ann Rogers and John Hill, "The principle of distinction is more clearly violated in the US practice of signature strikes that target unnamed/unidentified individuals simply because they exhibit a behavior the US considers threatening."[20] Because the operators do not know for certain whom they are attacking, they are operating in a gray area regarding discrimination—even though the target might exhibit many of the behaviors associated with belligerent status, they typically are attacked without warning in a noncombat situation. Analysts from RAND have argued that signature strikes fail to conform to the principles of Just War, in large part due to the unknown nature of the enemy being attacked.[21]

In theory, robotic sensors might prove more accurate than humans— already, facial recognition technology is proving extremely accurate, and it is likely to become more so in the near future. Yet, military robots are only as discriminating as their programming allows—and there may be significant military advantages derived from maintaining ambiguity and a "shoot first, ask questions later" mentality in autonomous robots' programming. Thus, it is impossible to determine if military robotics will have a greater propensity to engage in discrimination of targets than that demonstrated by their human operators to date. And, as Peter Asaro notes, "Discrimination is necessary but not sufficient for ethical killing in war."[22]

MILITARY NECESSITY

Necessity refers to the idea that warfare should not be conducted with naked violence as the primary goal but rather merely as a means to achieving a larger objective.[23] Some level of destruction is inevitable in warfare— otherwise, the conflict might be resolved through diplomacy, economic coercion, or some other means of interstate resolution. However, destruction for the sake of destruction is not an acceptable behavior. Targets that might be protected under the discrimination concept may become legitimate targets due to military necessity, under certain circumstances. Thus, a cultural heritage site that might ordinarily be immune from attack might become a legitimate target if it is used as a fighting position by an enemy military unit. One of the more famous examples came during World War II, when the ancient abbey of Monte Cassino, near Rome, was almost entirely destroyed by Allied bomber aircraft after being used as a defensive position by German troops. More modern examples have occurred when military units have used hospitals, schools, and religious buildings as fighting positions—each type of structure would normally be strictly off-limits but not if it is occupied by enemy fighters.

Even if a military action is both proportional and discriminate, it might still be immoral if it violates the principle of military necessity. Essentially, if there is no military gain to be made through taking a specific action, meaning no positive progress toward a belligerent's war aims, an action that will result in death or destruction should not be carried out. As with proportionality and discrimination, a few examples will help to flesh out this extremely important principle of morality in warfare.

In 1864, the U.S. Civil War entered its fourth year. Although the Union Army had slowly advanced into Confederate territory, it had done so at a steep price in blood and treasure, and the Confederacy showed few signs of imminent collapse. General Ulysses S. Grant, the newly named general in chief of the Union Army, devised a plan to attack the Confederacy on five separate fronts, hoping to stretch its defenses beyond the breaking point. Grant decided to accompany General George Meade's Army of the Potomac as it moved toward the Confederate capital of Richmond, Virginia. The advance provoked an inevitable clash with General Robert E. Lee's Army of Northern Virginia, which Grant hoped to either destroy or compel to surrender. Grant entrusted the second-largest campaign of 1864 to his closest lieutenant, General William T. Sherman.

Grant ordered Sherman to advance from southeastern Tennessee toward Atlanta, a key industrial hub and rail junction for the Confederacy. Grant surmised that the Confederates could not abandon the city without a fight and could not defeat Sherman's army in the field. Confederate general Joseph E. Johnston's Army of Tennessee mustered only 50,000 effectives, while Sherman's juggernaut advanced with over twice as many troops. Johnston used every delaying tactic and geographic advantage he could garner but simply did not have the manpower to halt Sherman's advance without losing his own army in the process. Despite his best efforts, his defenses were slowly pushed back into the outskirts of Atlanta by July. Frustrated by the gradual retreat, Confederate president Jefferson Davis replaced Johnston with General John Bell Hood, who elected to attack Sherman's larger force. This proved a disastrous decision that hastened the city's fall and the destruction of the Army of Tennessee.

Having captured Atlanta, or at least what was left of it after the fires set by retreating Confederates had died down, Sherman telegraphed news of his victory to Washington. He then determined that military necessity justified a decision to bring the war directly to the people of the South. To signal his intentions, Sherman ordered that the city of Atlanta be evacuated and even offered to provide the necessary transportation for civilians to flee the site of the war. However, when the mayor of Atlanta, James Calhoun, protested this decision, Sherman's written response demonstrated his vision of warfare and the concept of military necessity. On September 12, 1864, he wrote:

I have your letter of the 11th, in the nature of a petition to revoke my orders removing all the inhabitants from Atlanta. I have read it carefully, and give full credit to your statements of the distress that will be occasioned by it, and yet shall not revoke my orders, simply because my orders are not designed to meet the humanities of the case, but to prepare for the future struggles in which millions of people outside of Atlanta have a deep interest. We must have peace, not only at Atlanta but in all America. To secure this we must stop the war that now desolates our once happy and favored country. To stop war we must defeat the rebel armies that are arrayed against the laws and Constitution, which all must respect and obey. To defeat these armies we must prepare the way to reach them in their recesses provided with the arms and instruments which enable us to accomplish our purpose. Now, I know the vindictive nature of our enemy, and that we may have many years of military operations from this quarter, and therefore deem it wise and prudent to prepare in time. The use of Atlanta for warlike purposes is inconsistent with its character as a home for families. There will be no manufactures, commerce, or agriculture here for the maintenance of families, and sooner or later want will compel the inhabitants to go . . .

   You cannot qualify war in harsher terms than I will. War is cruelty and you cannot refine it, and those who brought war into our country deserve all the curses and maledictions a people can pour out. I know I had no hand in making this war, and I know I will make more sacrifices today than any of you to secure peace. But you cannot have peace and a division of our country.[24]

Sherman's forces commenced a march from Atlanta to Savannah, where they expected to be resupplied from the sea. On the way, Sherman ordered his troops to engage in a scorched earth campaign, tearing up railroad tracks, destroying factories, and burning plantations whose owners continued to support the Confederate cause. He did not turn his troops loose to kill and rape, but he did encourage ever-more effective means of property destruction. During his infamous "March to the Sea," Sherman's troops acted like a magnet for escaped slaves, whom he brought into his lines and employed as menial laborers to support his cause. To this day, Sherman remains vilified in much of Georgia and the Carolinas (where he proceeded next), regardless of the military necessity of his actions.[25]

   In 1864, the South American nation of Paraguay declared war upon Uruguay. It also chose to align itself against its far-larger and more powerful neighbors, Argentina and Brazil. The Paraguayan military commenced its war with over 60,000 troops, drawn from a population of 450,000. In comparison, their foes, the Triple Alliance, boasted fewer troops but a combined population of 11 million, more than 20 times larger than Paraguay's, and a massive economic and industrial advantage to boot. In short, Paraguay, despite being widely regarded as the aggressor, faced astronomical disadvantages in prosecuting its war. For their part, the Triple Alliance showed little interest in carrying on the war and even less interest in destroying Paraguay, despite the ease with which such an objective might have been accomplished. In many ways, the Triple Alliance showed almost inhuman restraint in trying to resolve the war without eradicating their tiny neighbor—but it was almost to no avail. The Paraguayan military insisted upon almost suicidal activity, eventually conscripting boys as young as 10 and men as old as 60. By the time the Triple Alliance finally forced Paraguay to accept a peace treaty, the aggressors had lost at least 70 percent of their adult male population. Those who remained alive did so largely due to the Triple Alliance's unwillingness to conduct military operations unless absolutely necessary and their determination to retain the moral high ground in an extremely asymmetrical conflict.

From the period 264 BCE to 146 BCE, the Roman Republic fought a series of wars against its primary rival in the Mediterranean: Carthage. The First Punic War (264–241 BCE) revolved largely around control of Sicily, a potentially valuable island situated directly between the warring capitals. When the Romans won the first war in 241, they imposed a significant indemnity and military restrictions upon Carthage. Undeterred, the North African city colonized Spain, and most of the Second Punic War (218–201) was fought on the Italian Peninsula. Once again, the Romans managed to triumph at enormous cost and imposed an even harsher peace upon their foes. To some Roman politicians, no peace short of complete conquest and subjugation of Carthage could guarantee Roman security. Senator Cato the Elder famously concluded many of his speeches, even those on completely unrelated subjects, with a call for the annihilation of Carthage. When the Third Punic War (149–146) erupted, it was as much a war of Roman conquest as a contest between rivals, and the relative wealth and populations of the two states all but guaranteed a rapid Roman victory.[26]

Roman general Scipio Aemilianus, the adopted grandson of the Second Punic War's victorious General Scipio Africanus, marched his forces directly from the sea to the city of Carthage. After a siege of three years, during which thousands of the city's inhabitants starved to death, he managed to compel the surrender of the remaining 50,000 citizens. Roman troops rounded up the survivors and immediately sold them into slavery. The city was systematically looted and burned for nearly three weeks, and then Roman engineers oversaw slave labor in the complete destruction of every building within the city.[27] Having eradicated the physical city and removed its population, the Romans next sought to eliminate almost all references to its existence, as if to remove it from recorded history. Undoubtedly, Scipio Aemilianus could have coerced the city to surrender by offering terms, but he preferred instead to use the city and its inhabitants as an object lesson for any civilization seeking to challenge the might of Rome. His actions moved far beyond any defensible notion of military necessity—but also stood as a lasting testament to the power of fear and rivalry.

During the German domination of the European continent at the height of World War II, German military and political leaders met at Wansee on January 20, 1942. At that conference, they discussed the advisability of conducting a genocidal campaign against the European Jewish population and other groups that the Nazi Party considered subhuman. One of the key figures of the Wansee Conference, SS Obergruppenfuhrer Reinhard Heydrich, presented what he termed the "Final Solution," a plan to track down and murder every European Jew. In addition to serving as Heinrich Himmler's deputy in the SS, Heydrich was also the protector of Bohemia and Moravia. On May 22, 1942, two British-trained Czechoslovakian rebels ambushed Heydrich's staff car and gravely wounded him. One week later, he died from his injuries, depriving the Nazis of one of their most effective, fervent, and diabolical leaders.

The German high command, infuriated by the assassination, traced the attackers back to the Czech villages of Lidice and Ležáky and determined to wreak an awful vengeance as a message to other occupied populations that might be considering open resistance. On June 10, 1942, German Army units entered the towns and rounded up the inhabitants. Every male citizen 16 and older was immediately executed, after which the towns were razed to the ground. The incident, commonly called the Lidice massacre, was "justified" by the Germans as a mechanism to halt a budding insurgency before it could spread, thus suggesting that military necessity might justify the horrors of the act they had just committed. That justification has had little effect upon the historical discussions of the incident and of German wartime behavior in general, as the massacre was just one of a number of mass atrocities perpetrated by the Germans upon defenseless civilian populations.

Military robots make the application of force much easier for a belligerent, particularly in terms of domestic politics. When one can make war with little or no risk to one's own forces, it becomes easier to internally justify the use of force, and the standard of military necessity potentially declines. Armin Krishnan, one of the foremost writers of the ethics of military robotics, notes: "I am concerned about using robotic weapons against terrorists because it makes it so easy for the armed forces and intelligence services to kill particular individuals, who may be guilty of serious crimes or not."[28]

THE MORALITY OF ROBOTS

Military robots, to date, are not programmed with any degree of moral governor.[29] With the exception of strictly defensive platforms, though, they still have a human being ultimately responsible for making the decision to take a life. This consideration is important, as many leading scholars believe that military robots, and in particular drone strikes, are eroding the morality of the nation that employs them. For example, Medea Benjamin argues that "from Pakistan to Yemen to Gaza, drone warfare snuffs out the lives of innocent civilians with impunity and renders thousands more maimed psychologically, left homeless and without livelihoods. In the name of the war on terror, drone warfare terrorizes entire populations and represents one of the greatest travesties of justice in our age."[30] In her estimation, it is impossible to use such platforms in a discriminate and proportionate fashion—and the U.S. military and intelligence agencies do little to even attempt restraint on moral grounds. Phyllis Bennis underscores this concept, on the grounds that killing has become the response of first, rather than last, resort. She argues that the United States is on a slippery slope as succeeding administrations continue and expand the policies of their predecessors, noting, "During the Bush years, when Afghanistan was kept in a distant second place to the priorities of the Iraq war, US policy on Afghanistan had included assassination. But it was during the Obama years that assassination in US war strategy became the tactic of choice."[31]

To Michael Walzer, one of the foremost modern theorists of Just War, it is simply impossible to engage in a war without risk to one side and to fight that conflict in a just fashion. As he summarizes his own position, "This is not a possible moral position. You can't kill unless you are prepared to die."[32] The ability to wage war with impunity is, therefore, an inherently immoral position. Assumptions about the nature of military robotics, including their capabilities and ability to fight in an ethical fashion, are helping to drive the assumption among many American citizens that drones offer an ethical means of conducting modern warfare with minimal casualties on each side. To Grégoire Chamayou, "The more widespread the legend of the ethical robot becomes, the faster the moral barriers to the deployment of killer robots give way. One might almost forget that the surest way to make the potential crimes of the cyborgs of the future impossible is still to kill them immediately, while they are as yet unhatched and there is still time to do so."[33]

Military robots might actually offer the greatest possible solution to the problem of military necessity, if they are required to utilize nonlethal means of disarming or incapacitating an enemy. Because robots are inherently expendable when compared to humans, they might be expected to undertake more risks to neutralize enemies without killing them. By not placing one's own forces in danger, it becomes far easier to operate with restraint and patience, particularly if the enemy in question has no means to leave the immediate area and essentially escape the region. It is a fair speculation to suggest that military robots should be limited to only nonlethal means, although that might be an impossible standard to expect. John Canning believes the future of military robotics should be an extremely advanced capability to disarm enemies without offering harm to the humans.[34]

Because machine logic does not always follow human reasoning (which is often shorthanded as "common sense"), programming military robots to behave in a moral fashion may be a nearly impossible task.[35] Several companies are currently considering the possibility of using flying robots to deliver small packages to consumers—but they are struggling with how to handle the "unknowns" of the flying environment. Further, once the robot reaches the desired location, there are still a host of other potential issues at hand. Consider a normal package delivery to a residential address—the driver arrives at the correct location and approaches the front door. The next step is to ring the doorbell, or if there is no such bell, to knock. If there is no answer, most delivery personnel will seek to place the package in a semiprotected location where it will be obvious to the residents upon their return. This requires a certain degree of human judgment—should it be left on the front doorstep? The driveway? Inside the back fence? Without a human operator, the envisioned delivery robots might be forced to choose between "deliver" and "return to depot," with nothing in between. This hardly makes for an efficient system—and it barely scratches the surface of the number of judgments that need to be made for deliveries that seem obvious to humans. If the delivery is done by GPS coordinates, rather than street addresses (a likely solution), it is entirely likely that there will be a rash of complaints about packages left on rooftops, in trees, and in other inaccessible, but technically correct, locations.

As is the case with the laws of war, military robots do not inherently change the nature of warfare and hence do not necessarily alter the ethics of conducting a just war. However, they certainly enable extremes in behavior, either to the positive or to the negative side of the behavioral spectrum. Ironically, they might make the conduct of wars in an ethical fashion more likely while also making the commencement of war a higher probability—leading the reader to wonder if they are a net gain or loss in terms of the morality of warfare. Or, to put it another way,

Today, as drone technology shields soldiers and citizenry from the costs of war, they must face the disturbing prospect of having greater latitude in the strategic choices that they make and condone, even if those choices run counter to long-standing legal or ethical norms. Drone warfare makes fighting just wars easier (and this is unquestionably good), but it will also facilitate the prosecution of unjust wars. Additionally, over reliance on military robotics may tempt individuals to confuse just and unjust wars.[36]

Lethal robots might be eroding our morality by making it far easier to contemplate killing our way out of political messes rather than seeking alternate resolutions—they might essentially make war the first, rather than the last resort, of modern nations.

CHAPTER 7

The Global Competition

Never despise your enemy, whoever he is. Try to find out about his weapons and means, how he uses them and fights. Research into his strengths and weaknesses.

—Aleksander V. Suvorov, 1789

Although the United States has been the earliest and most enthusiastic adopter of military robotics, it is certainly not the only nation moving into this new field of warfare. Not only are other nations examining the behavior of the American military (and the reaction of the international community), they are developing their own approaches to automating the battlefields of the 21st century. While a full examination of every nation active in this movement is well beyond the scope of this work, it is certainly worthwhile to spend time looking at the near-peer competitors of the United States, as well as some of the actors pioneering military robotics in different directions. To that end, what follows is an examination of a handful of nations that have adopted alternate strategies regarding the weaponization of machines in military operations. Each has developed sophisticated capabilities in either the cyber or robotics domains, sometimes both, but has done so with different priorities and expectations than the United States.

One of the mechanisms by which modern nation-states can and do conduct all manner of hostile actions against one another is through the cyber domain. Like military robotics, cyber warfare is entirely dependent upon the development of advanced machines, which are in turn dependent upon ever-increasing computer sophistication. Attribution of cyberattacks is one of the most vexing problems of cyber warfare, and it is not always easy to clarify responsibility for activities conducted via remotely piloted vehicles or autonomous machines.

Military robotics is inextricably linked to advances in computer technology, and thus it is only natural that any nation attempting to shift the burdens of warfare away from its human population and onto machines would wish to pursue the most advanced cyber and computing capabilities. It is unsurprising that the nations with the most advanced military robotics also possess the most advanced cyber capabilities, although some degree of specialization in both the design and employment of these machines has proven almost inevitable. How a nation uses its cyber and robotic assets tends to mirror how it has traditionally utilized its human military and nonmilitary resources—in most cases, nations have a preferred approach to international conflict that guides acquisition programs, patterns of strategy, and the development of doctrine. How a nation behaves in cyberspace, in particular, tends to mirror how that nation acts in the physical realm. Thus, a nation that prefers industrial espionage to internal innovation will likely follow the same pattern through computer networks wherever possible. A nation that prefers to rely upon massed wave attacks and brute force on the battlefield rather than employing an indirect approach is likely to favor the same methods in cyberspace.

So far in the 21st century, cyber activity has primarily served as a secondary effort and enabler for military operations, although in some cases cyberattacks have managed to achieve effects comparable to those sought through kinetic attacks but without the corresponding collateral damage. Cyber operations have proven especially adept at espionage activities, as the data-driven cyber domain is expressly built for the rapid transmission of enormous volumes of information. A single portable hard drive that can easily fit in the palm of one hand can currently store more information than an entire university library of bound volumes held 20 years ago, and that information can be quickly and easily copied, transferred, modified, or discarded. Yet espionage is not warfare, nor has it ever been the primary cause of a major war, even if skilled cyber spies might be able to infiltrate secure servers and steal billions of dollars' worth of information. Cyber security company McAfee estimates that cyber espionage costs U.S. companies more than US$100 billion per year, with global losses expected to reach US$2 trillion by 2019.[1]

Cybercrime is another form of cyberattack that resembles cyber warfare but with a different objective. Typically, cybercrimes are committed for material gain, and they exploit the nature of the cyber domain by intercepting or stealing data and using it to engage in illicit financial transactions. The largest number of cybercrime attacks in the world originate from servers in the United States, although it is difficult to determine if they are actually being launched by American actors or if they are simply relying upon lax security in proximity to their targets. Cyber criminals target individuals, groups, and corporations, usually on an opportunistic basis. The most lucrative cyberattacks in history have involved breaking into corporate servers and stealing either large amounts of personal information to facilitate identity theft or enormous volumes of credit card transaction data and using it to transfer funds from hijacked accounts.

Cybercrime has steadily grown throughout the past two decades as computers and online financial transactions became ubiquitous. The United States is the nation where the largest percentage of mass attacks originate. These consist of unsophisticated attempts to find victims via mass emailing of fraudulent messages, such as the infamous Nigerian prince scam. More sophisticated attacks often originate outside of the United States, in large part because international attacks are harder to trace and capturing cyber criminals across international boundaries requires the active cooperation of potentially hostile foreign governments. A number of international cooperatives have emerged, through which cyber criminals can traffic in stolen instruments, exchange successful hacking programs and techniques, and even rent or sell botnets. The largest and most notorious such cooperative is the Russian Business Network (RBN), which initially served to anonymize server traffic and thus facilitate attacks but which has grown into a multinational criminal organization capable of launching extremely sophisticated attacks. The RBN is at least protected by the Russian government, which adamantly refuses to allow any law enforcement or cyber security investigators to have access to the organization or its leadership.

RUSSIA: RESURGENCE OF THE BEAR

Russia has a strange civil-military partnership when it comes to international conflict. In particular, the Russian intelligence services seem to rely very heavily upon virtual cyber militias for offensive cyber actions, particularly for harassment campaigns against less-capable cyber foes. These groups tend toward brute-force cyberattacks rather than any particularly sophisticated approaches, but that does not negate their inherent effectiveness. Also, because they are not directly tied to the government, they offer a certain deniability to the Russian political leadership, should events go awry or spin out of control. In the 21st century, Russia has proven to be a particularly difficult neighbor, engaging in conflicts with Estonia, Georgia, and Ukraine through both the cyber and physical domains. Military machines have offset some of Russia's personnel problems, allowing the government to target weaker neighbors with a high degree of impunity.

Russia has spent decades trying to regain its position as a world superpower, but it struggles with geographic, demographic, and economic problems. When the Soviet Union collapsed in the early 1990s, many of its separate republics seized the opportunity to declare independence from the emerging Russian Federation, even though most had been under the yoke of Russian domination when the Soviet Union formed. Russian patriots tended to see these independence movements as a betrayal of the greater good, the actions of ungrateful citizens, as each defection weakened the remaining state. For their part, the newly independent states saw this as an opportunity to reclaim national and ethnic heritage and to emerge from decades or centuries of Russian oppression. Many had attempted to cast off Russian control during the Russian civil war (1918–1921) or during World War II. Thus, when another opportunity to proclaim independence arose, it was almost inevitable that a number of aspiring states would take the risk of war with Russia to emerge from the ashes of the Soviet collapse as independent nations. With each defection, Russia's potential manpower reserve dwindled, and by the year 2000 it had less than half as many citizens as the United States. Yet, it still is the largest nation on earth and not surprisingly has the longest hostile borders to defend—making the low population and depressed economy more glaring threats.

None of the small states around the Russian periphery could realistically expect to declare independence and hold it through force of arms if the Russian government decided to strongly contest the matter. Their independence relied in large part on Russian willingness to let them leave, either because the Russian government sought to appear as a benevolent member of global society that did not wish to hold territory by force or because it was distracted by other issues and did not choose to reinvade its neighbors. For the first decade after the Soviet collapse, the Russian government proved relatively willing to allow states to proclaim their independence, with the significant exception of Chechnya, where the Russians inexplicably chose to take a stand against a breakaway move. However, several of the new states realized that their best guarantee of continued independence came not from relying upon Russian benevolence but from securing the protection of powerful partners that might view Russia as a rival. In the Baltics, protection came in 2004 in the form of admittance to the European Union and the North Atlantic Treaty Organization (NATO), a decision that followed the path set by former Warsaw Pact members who had never been formally annexed into the Soviet Union but which had lived under its direct and indirect control throughout the Cold War.

Although NATO was willing to induct Estonia, Latvia, and Lithuania as new members, there was some question of how it might defend the small states in the face of naked Russian aggression. None of the Baltics possesses a large population or a formidable military; in the event of a Russian attack, all three would likely be completely overrun long before anything more than a token NATO force could come to their assistance. Yet, even such a token force might serve as a sufficient deterrent given that any attack against it might provoke a massive reprisal against an aggressor state. In some fashions, the same system that had hedged against Soviet aggression in Europe throughout the Cold War had been redeployed against Russia, with only the location of the front lines having changed. Undoubtedly, Russian leaders understood that NATO now effectively functioned as an anti-Russian alliance, regardless of any activities NATO might undertake in the Balkans or elsewhere.

Surely, the Russians chafed at the feeling of being hemmed in by a rival that had been an open enemy a short time earlier. The cheekiness of the new NATO members exacerbated the situation, as they seemed to be openly taunting their former overlords by hosting exercises and demonstrations and reveling in their new security under the American military umbrella. In 2007, the Estonian parliament passed the Foreign Structures Law, an edict that required any Soviet-era monuments or similar structures to be removed from public lands. The Estonian government was demonstrating in a clear fashion that it did not look fondly upon its years as a Soviet republic and that any vestiges of the time period should be removed to eliminate reminders of the era. The Russians reacted like a surprised spouse receiving a divorce decree—a combination of shock and rage swept through Russia, given that the Estonians could so casually cast off a decades-long relationship, a shared history of sacrifice and struggle between the two peoples. When the Estonians decided to move a massive bronze statue of a Soviet soldier from the heart of the Estonian capital, Talinn, to a more remote military cemetery, the ethnic Russian minority was furious. They saw the uprooting of the statue as disrespect for the sacrifices of Soviet troops that liberated Estonia from German occupation, and they loudly voiced their displeasure to their contacts in Russia.

What followed was not an overt attack, such as an invasion of an armed force marching across the border in an act of naked violence. It might not have even occurred at the instigation of the Russian government or involved any state resources. It did no permanent physical damage to any Estonian assets and did not wound, much less kill, any Estonian troops. Yet, it was still a devastating demonstration of how modern machines can be used to carry out aggression. When the Russians decided to unleash their anger on Estonia, they did so through the largest coordinated cyberattack that had ever been launched. Millions of computers based in Russia or under the control of Russian botnets commenced a distributed denial of service attack against Estonian government and economic infrastructure targets. Essentially, these attacking computers began to bombard Estonian Web sites and servers with simple, automated requests for information. As more and more computers joined in the attack, their relentless queries overwhelmed the Estonian systems' abilities to respond, triggering a shutdown of server traffic and essentially bringing the nation's Internet service to a complete standstill.

For some nations, losing Internet service would be an irritant, not a catastrophe. Just as the occasional power outage might make life unpleasant and even indirectly cause harm or injuries, losing cyber functions might be akin to a temporary utility outage. This was not the case in Estonia in 2007—the nation was one of the most networked and heavily

Internet-dependent countries on earth. Estonia had become an almost cashless society, relying on electronic fund transfers to complete even the most basic transactions for goods and services. Losing the ability to use Internet connections brought the small Baltic nation's entire economy to the point of collapse. Shortages soon emerged, because the nation's merchants relied upon just-in-time shipments to keep their shelves stocked. Citizens attempted to withdraw cash from the Estonian banking system, only to discover that the banks required Internet connections to verify account balances, and thus the banks proved unwilling to distribute their small cash reserves.

Estonia appealed to NATO for assistance with the crisis, on the grounds that it was under attack from a foreign source. Although NATO cyber experts attempted to mitigate the attacks, the full NATO council decided that Estonia's situation did not rise to the level of an armed attack and thus decided that no direct response from the alliance was merited by the cyberattacks. When cyber forensics specialists attempted to trace the source of the attacks, they discovered to no one's surprise that most of them originated in Russia or travelled through Russian servers. Also unsurprisingly, the Russian government adamantly denied all responsibility for the attacks and absolutely refused to even consider allowing NATO cyber experts access to Russian computer networks. Despite Russian intransigence, cyber security professionals determined that some of the code used in the attacks had been written on computers employing Cyrillic character keyboards and almost all of the attacks came from botnets controlled from Russian territory. When confronted with this evidence, Russian government officials suggested that the attacks were the activity of patriotic hackers outside of the government's control and that the Russian law enforcement bureaucracy lacked the legal rights and abilities to trace and halt the activities of its nationalist citizens.

NATO's refusal to treat the Estonian cyber crisis as a deliberate attack undoubtedly emboldened the Russian government to increase its reliance upon cyber warfare as a means of leveraging influence over its neighbors. Although NATO soon created a Cyber Defense Center of Excellence in Talinn and commenced collective cyber activities at the new site in 2008, it did little else of substance to stop the attacks. Attempts to directly counter the cyberattacks had little effect: it was soon discovered that the botnets were being continually reprogrammed and their inquiries rerouted through new servers to keep up the intensity of the attacks. This behavior suggested that a central controller was directing the cyber campaign and taking steps to maintain its effects. The central control theory was bolstered by the abrupt end to the attacks, which suddenly ceased after causing three weeks of chaos. Having proved their point, the Russians seemingly decided that they had inflicted enough punishment upon their neighbor for the impertinence of moving a statue. Had the attacks been the result of spontaneous mob activity, they would have likely receded gradually, rather than in such a precipitous fashion. Throughout the period of attacks, the work of moving the statue continued as planned. It now resides at a national military cemetery, although the Estonian government did underwrite some previously unplanned beautification projects around the statue's new location, a decision that perhaps mitigated the insult of moving it in the first place.

Russia certainly recognized the devastating effect a coordinated series of cyberattacks can have upon a population heavily dependent upon computer networks, particularly if those networks were established with efficiency and convenience as far more important considerations than security. Even if the unsophisticated attacks had little chance of disrupting military cyber systems, they still offered a means to coerce an enemy civilian populace, and, by definition, disrupt military activities due to the heavy dependence of Western militaries upon civilian augmentation. Thus, it is unsurprising that when the Russian government engaged in another confrontation with a former Soviet republic, it would use the same cyber technique for coercive purposes. In 2008, ethnic Russian citizens of Georgia living in the semiautonomous provinces of South Ossetia and Abkhazia demanded independence so that they could petition for Russian annexation. Faced with the loss of a significant portion of its territory, Georgia refused to acquiesce to the demands. The Russian government claimed that ethnic Russians faced discriminatory policies from the Georgian government and elected to send military forces to the region on the grounds that it had a responsibility to protect ethnic Russians wherever they might live.[2] In any event, the small Georgian military sought to defend its borders against Russian incursions, a necessary exercise for any sovereign nation, but a hopeless one in this case. Not surprisingly, the Russian troops, backed by Russian airpower, had little difficulty in defeating the Georgian forces. Georgia had applied for NATO membership in 2005, but its application had been rejected by the alliance due to concerns about the autocratic government and the rampant corruption that supported it. Thus, Georgia found little support in the international community beyond weak protests regarding violations of national sovereignty. Russia claimed it was only engaging in humanitarian activity, seeking to prevent the Georgian government from murdering elements of its own population. Finding its entire territory under threat of invasion and conquest, with no international assistance forthcoming, the Georgian government capitulated and renounced claims to the breakaway republic.

What many Western observers of the Russia-Georgia conflict did not realize at the time is that Russian hackers had once again swung into action, unleashing millions of computers in a coordinated attack upon Georgian cyber networks. In a strange twist, Russian programmers wrote an application entitled Georbot that enlisted computer users from around the world to participate in the attacks. Infecting machines with malware to make them part of a botnet controlled in Russia was nothing new—the practice had occurred for nearly two decades prior to the debut of Georbot. What made this program unique was that its creators openly advertised the purpose of the software and convinced hundreds of thousands of computer users to deliberately infect their own systems with a virus that caused them to relinquish control over their own computers. Cyber security analysts discovered that Georbot was voluntarily downloaded onto computers so that the computers' owners, many likely patriotic Russian expatriates (or ardent University of Florida football fans mistaken over which Georgia was being targeted), could feel they were making a contribution to the already lopsided struggle.

When Russian leader Vladimir Putin began complaining that ethnic Russian citizens of Ukraine were being mistreated by their government, it was a sure sign that aggression against yet another former Soviet republic would soon commence. In 2014, Putin suggested that the Crimean Peninsula, home to a major Russian naval base and a Russian ethnic majority, should be ceded to the Russian Federation. The fact that the Ukrainian government had considered cancelling the Russian naval lease at Sevastopol was written off as merely a coincidence by Putin and not as a triggering event. Ukrainian protests fell on deaf ears, both in Russia and the United Nations. A Crimean plebiscite marred by accusations of fraud voted to leave Ukraine and join Russia, and, shortly afterward, Russian troops entered the region to secure the bloodless conquest. Because the peninsula is connected to the continent within the territorial borders of Ukraine, this created an unstable situation in which all supplies to the area would have to be shipped on the Black Sea or flown in by cargo transport. To alleviate the problem, Russia began demanding territory in the eastern portion of Ukraine, as well as transit rights. Soon well-armed, Russian speaking "rebels" began seizing positions in the area in an apparent breakaway attempt.

There was little the Ukrainian government or military could do to recover Crimea—the Russian fait accompli had moved far too quickly to be stopped by the minimal defenses on the peninsula. Protests to the United Nations did not result in any direct military assistance, as the Russian Federation's delegation could simply veto any resolution of the UN Security Council to use force in the region. As was the case a few years earlier in Georgia, Russian aggression proved successful enough to seize territory at the cost of a few poorly enforced, short-lived economic sanctions, hardly enough to temper the ambitions of a resurgent Russian empire. Also similar to the Georgia situation, Russian forces relied upon massive cyberattacks as a means to disrupt military communications, hinder effective command and control of Ukrainian fielded forces, and introduce confusion into the logistics system supporting those forces. Cyberattacks also targeted Ukrainian government agencies, major sectors of the economy, and even social media operating in Ukraine. Once again, the Russian government disavowed all knowledge of the attacks while suggesting that Russian patriots might have taken exception to media reports of Russian attacks in Ukraine and decided to fight back through the cyber domain.

In these Russian cases, cyberattacks served as the primary form of aggression against Estonia, a NATO member, and a secondary form of aggression against Ukraine and Georgia, both of whom had unsuccessfully petitioned for NATO membership. This suggests that Russia is willing to engage in overt military action when it is confident that it can control any potential escalation, while it prefers to remain covert if there is the possibility of intervention by an outside force that might offer a significant threat to Russian security. Each of these attacks demonstrate how even an incredibly simple form of cyberattack can be very effective if it is conducted upon a large-enough scale, particularly if the primary objective is to deny the use of cyberspace to the enemy. Distributed denial of service attacks are the most basic and unsophisticated of cyberattacks, but, as Senator Sam Nunn noted regarding a different set of circumstances, "At some point, numbers do count. At some point, technology fails to offset mass."[3]

In 2015, a Freedom of Information Act lawsuit by Judicial Watch seeking emails from former U.S. secretary of state Hillary Clinton's time in office that pertained to the Benghazi consulate attack of September 11, 2012, inadvertently revealed that Secretary Clinton had conducted all of her State Department business outside of the department's secure network. Rather than follow established protocol as mandated by State Department policy and federal nondisclosure laws, Clinton chose to retain her communications on a server placed in her home in Chappaquiddick, New York. Clinton later claimed that she made this decision for personal convenience and that no classified information had transited the network. Her detractors claimed she had made this decision as a means to hide unethical connections between the State Department and the Clinton Global Initiative, a private foundation with an enormous list of high-powered donors. Regardless of Clinton's reasoning, setting up an email server outside of the government's secure network, which itself has not always proven completely secure, definitely opened her communications to attack and likely interception by foreign governments, and classified information at the highest level of sensitivity was definitely sent through her email server. On January 22, 2014, agents arrested Marcel Lehel, a Romanian hacker better known by his screen name, "Guccifer." Lehel claimed that he had infiltrated the Clinton server on at least a dozen occasions. While details of what Lehel actually found on the server have not been released, it is almost impossible to imagine that an international criminal who made millions of dollars through the theft and resale of personal information would fail to take advantage of the opportunity to sell access to the private communications of a sitting U.S. secretary of state. How much would a competing government pay for such access? And how many nations might pay millions for information contained within her communications? Details are beginning to emerge that suggest that at the very least, the Russian government used Guccifer's methods and point of access to download Clinton's files. These communications likely clarified the U.S. intentions toward the newly expansionist Russia and helped Vladimir Putin determine precisely how far he could push the United States and NATO without fear of an adverse response. The messages also contained lists of classified operatives stationed in undercover positions around the world, including in the U.S. embassy in Moscow.

Although the Russians have placed more emphasis and faith into the cyber domain than into the development of military robotics, there are still a number of remotely operated platforms of note within the Russian arsenal. Major weapons corporations in Russia have developed unmanned versions of their manned aircraft, leading to remotely piloted aircraft that look remarkably similar to MiG aircraft and Mil helicopters.[4] Such an approach means that there will be no need to create special-purpose facilities to utilize these systems, as they can operate from preexisting airbases and runways and can typically use the same type of fuel and armaments as their manned cousins. This is one mechanism to offset a chronic pilot shortage and retain operational capabilities that the nation otherwise might not be able to afford.[5]

CHINA: THE DRAGON AWAKES

By far, the largest and most successful perpetrator of cyber espionage against the United States is the People's Republic of China, which has used cyberattacks to steal unprecedented amounts of information from government agencies, major corporations, and private citizens. Much of Chinese cyber espionage has been targeted against companies and agencies associated with military innovation, a decision that has enabled significant advances in Chinese technological development. Observers have noted a decided similarity between the U.S.-built Joint Strike Fighter (F-35) and the Chinese J-31. Many Chinese unmanned platforms look almost identical to American counterparts, although it is impossible to determine if they have the same level of capabilities.

This brazen display of stolen technology essentially admits to, and even crows about, the theft of design parameters and construction specifications of America's most advanced manned military aircraft. Yet, the United States has launched only tepid protests of Chinese cyber espionage, even when major media outlets have published extensive coverage of these activities.[6] This may be due to the fact that China is the largest foreign holder of U.S. government debt. If the U.S. government pushes China too hard on espionage, China may in turn choose to call in its loans, a move that could trigger financial turmoil in U.S. markets. It may also be the case that the United States chooses not to protest too vociferously because its cyber agencies are equally guilty of plundering information from Chinese entities, and thus each side might be quietly tolerating the other's intrusions. However, if the United States is the world's chief innovator in military technology, it is unlikely to garner the same benefit from examining Chinese weapons technology as it gives up by tolerating intrusions. Also, the Chinese government tends to partner very closely with Chinese corporations and seems to provide information of interest to those companies that can help them compete with Western companies operating in the same economic sectors.

China seems to have a greater interest and capability in cyber security, demonstrated in large part by its ability to sever the Chinese Internet from the rest of the World Wide Web. The Chinese government has insisted that major computer corporations that wish access to the Chinese marketplace turn over source code for their most important products for inspection before they will be allowed to reach Chinese consumers. Most, though not all, major software companies have assented at least in part to these demands. Chinese censors have also demanded that major Internet search providers, such as Google, deliberately censor the results that they provide to users operating computers on Chinese soil. Resistance to this demand has led to a major conflict between Google and the Chinese government, which openly assists the largest domestic Chinese competitor to the American search engine giant.

Chinese espionage has allowed them to close the gap in technological capabilities but stealing information can never allow one to truly overtake a rival—nor does it build the internal capacity for major innovations. The Chinese decision to pursue high-technology military capabilities indicates a shift from the traditional Chinese way of war—reliance upon massive amounts of manpower to offset other disadvantages. This might indicate that the Chinese government is less willing to throw away the lives of its citizens in future conflicts or it might simply be a recognition that allowing too much disparity in technological capabilities is a recipe for disaster. After all, the numerical advantages of the Aztecs and Incas did nothing to offset the technology of the invading Spaniards, who were essentially fighting a riskless war in their conquests. It is also possible that China will turn its massive manufacturing capabilities toward the production of cheap military robots and then utilize them along their classical strategic lines, in overwhelming numbers against a technologically superior enemy that still cannot withstand the onslaught of millions of attackers.[7]

Politically, China seems to be dedicated to a very defensive mind-set, especially within its own region. The Chinese military has not developed major power-projection capabilities and seems instead to focus upon extending the perimeter of its defenses, particularly into the South China Sea.[8] The Chinese government has ordered the construction of a series of artificial islands in the region, each duly equipped with a small airfield. The islands also offer an unsinkable air defense platform for advanced long-range missile systems. If enough interlocking fields of fire are constructed, it might become cost prohibitive for the United States to attempt any form of incursion in the region, even with unmanned platforms. China has also created a large fleet of underwater unmanned vehicles, which might be used for clearing sea lanes, detecting submarines, planting naval mines, or launching direct attacks against enemy ships.

In some fashions, machine warfare might offset the disadvantages a nation possesses against stronger neighbors. For example, a nation with a comparatively small population base but a large capacity for industrial production might choose to churn out enormous numbers of semiautonomous robots for combat. This approach might allow Taiwan, which is heavily industrialized but geographically limited by the size of its island, to create a deterrent against invasion from the Chinese mainland. The People's Republic of China possesses a population approximately 60 times larger than Taiwan's and could surely mount an overwhelming invasion force should it decide to do so.

For the past seven decades, Taiwan has relied upon the U.S. Navy as the key to its protection, and, in turn, the U.S. Navy has used its fleets to deter such an invasion. There is no guarantee that American foreign policy will continue to protect the island. There is also some question as to the U.S. military's willingness and ability to fight against a near-peer competitor so close to that rival's home territory—any battles over Taiwan would be fought within range of Chinese airbases and missile batteries, while American forces would be almost entirely dependent upon carrier-based airplanes in the immediate region and long-range strike aircraft from comparatively distant bases. If the Chinese government decided to test the mettle of the United States, it could hardly choose a more advantageous location, and all of the potential participants know it. As a result, Taiwan needs to maintain enough of a deterrent to hold off Chinese aggression, without turning itself into an open threat that might accidentally provoke a Chinese attack. For this reason, nuclear weapons, arguably the most powerful of deterrents, do not make strategic sense for Taiwan, while the construction of armed robots might be more successful. Taiwanese companies have begun purchasing, designing, and building unmanned military vehicles as quickly as possible, and they are being fielded for coastal patrol, to provide emergency communications nodes and to offset numerical disparities.

ISRAEL: SURROUNDED BY ENEMIES

The state of Israel has a long history of contentious relations with its neighbors. When David Ben-Gurion proclaimed the independent state of Israel on May 15, 1948, the move provoked an immediate attack by five neighboring nations, augmented by volunteers from several more. After more than a year of bloody fighting, the nascent United Nations finally managed to broker a very shaky cease-fire. In addition to the almost-continual border skirmishes, Israel fought a pair of major wars against Arab coalitions in the three decades that followed. In each case, Israel survived the best efforts of its antagonists and even managed to capture and hold key defensible terrain. After the 1973 war, the Egyptian government of Anwar Sadat agreed to directly negotiate with the Israelis at Camp David, signing an accord with the Jewish state in 1978. The other hostile neighbors have steadfastly refused to negotiate with Israel or to recognize its right to exist. As a result, Israel has always pursued a strong military capability, one that is able to offset the massive numerical advantage of its adversaries. In particular, Israel has depended upon maintaining a significant technological edge over its competitors, both through the import of advanced weaponry and through the development of a domestic research and production capability that avoids complete dependence upon outside assistance. Israeli scientists and engineers have developed nuclear energy facilities and certainly have both the knowledge and the capability to produce nuclear weapons. Many intelligence agencies believe the Israelis have built at least a small arsenal of nuclear weapons to be used as a last resort if the nation's conventional defenses are on the verge of collapse. For its part, the Israeli government has never confirmed or denied such a stockpile, preferring instead to pursue a deliberate ambiguity that might deter a potential adversary more effectively than the sure knowledge of Israeli nuclear capabilities.

In addition to maintaining a strong military, the Israeli government has always been extremely aggressive toward any external threats. This has manifested itself through a very robust espionage program, a commitment to near-universal military service for the citizenry, and a willingness to engage in preemptive attacks if they are deemed necessary to the survival of the state. In 1967, when Israeli agents and border observers detected a military buildup in Egypt and Syria (which at the time operated under a shared government dubbed the United Arab Republic), the Israeli Air Force launched a spoiling attack against the Egyptian Air Force. The strikes caught the Egyptians completely unprepared and virtually destroyed their entire air force at minimal cost to the Israelis. The next day, a similar attack almost completely destroyed the Syrian Air Force. Without air cover, the advancing Arab armies, which made the ill-fated decision to continue their assault without regard for the aerial situation, proved easy targets for airstrikes, and the Israelis won a decisive victory in the Six Day War. The ease of their triumph led to overconfidence, such that Israel Defense Force (IDF) commanders assumed such an attack could be replicated in the event of future impending conflict.

Arab commanders had good reason to fear future aerial attacks from the IDF but also reasons for optimism. To take advantage of Israeli hubris, they purchased extremely sophisticated integrated air defense systems (IADS). By 1973, the radar stations, artillery, and missile batteries were emplaced and ready to repel any inbound aircraft. Arab armies openly mobilized in such a fashion that the Israelis could not miss the signs of an impending invasion. Again, key leaders in the Israel Air Force (IAF) recommended a preemptive strike, but this time Prime Minister Golda Meir chose a defensive strategy of allowing the enemy to make the first move. She felt that the world would blame Israel for the conflict if the Israelis launched the first attacks of a second war in six years. When the IAF finally assumed the offensive, it flew directly into a trap, attacking state-of-the-art defenses that took a frightful toll of Israeli aircraft. When the Israelis sought to offset their aerial losses through the advance of armored formations on the ground, they discovered that their Arab opponents had also purchased advanced antitank missiles to offset the Israeli technological advantages in ground warfare. The 1973 War soon devolved into a brutal slugging match—the Israelis had the advantage of fighting a defensive conflict from prepared positions, but the Arabs had the weight of numbers. The Israelis maintained air superiority but could not utilize their airpower advantages due to the enemy's IADS. In turn, the Arabs were relatively safe under the umbrella of their IADS but could not advance beyond the limit of its coverage, and they could not move its elements forward without exposing their entire IADS to destruction. A stalemate ensured, and eventually armistice negotiators once again managed to craft a cease-fire. Israel emerged from the conflict determined to maintain a major technological advantage over its rivals, by any means necessary, and to increase domestic capability for producing military hardware. It also decided to pursue an even-closer relationship with the United States, which had transferred an emergency shipment of combat aircraft to the beleaguered Israelis through Operation Nickel Grass. The two nations soon began to cooperate in weapons development programs, including the Patriot missile system and a series of remotely piloted aircraft.

The Israeli government feared that if any of its enemies managed to reach parity in military technology, Israel might be overwhelmed before any external support could arrive. In particular, they feared that an Arab enemy might successfully undertake a nuclear weapons program and hence be able to place the entire Jewish nation at risk. Iraq, which had participated in the 1948 and 1967 wars but remained out of the 1973 conflict, commenced a nuclear weapons program in the late 1970s or early 1980s. President Saddam Hussein believed the possession of nuclear weapons would confer upon him the mantle of leader of the Arab states of the Middle East. It might also facilitate emerging victorious from the Iran-Iraq War, which commenced in 1980 and soon turned into a no-holds-barred struggle involving the use of chemical weapons and child soldiers. Even if the Iraqis did not choose to launch nuclear attacks upon Iran, their capability to do so might force the Iranians to concede Iraqi territorial gains in the war.

Even with the Iraqis engaged against the Iranians, the Israeli government could not countenance the idea of a nuclear-armed Iraq, and it commenced planning an airstrike to destroy the Iraqi nuclear reactor at Osirak, the home of the nuclear weapons program. On June 7, 1981, IAF warplanes launched an attack against Osirak, annihilating the Iraqi nuclear program and clearly demonstrating that Israel would not tolerate nuclear weapons in the hands of a state that had repeatedly attacked Israel. To avoid alerting the Iraqis to their intent, much of the airstrike's flight path was over Saudi airspace. To this day, some controversy exists over whether the Saudis knew of the planned airstrike and allowed it to go forward for their own ends, whether the Saudis detected the overflight but chose not to intercept it out of fear of Israeli capabilities, or whether the Saudi air defenses completely failed to detect the attacking warplanes as they transited Saudi airspace. Although the Saudi government had repeatedly denounced Israel, it had not directly participated in any of the previous wars against Israel. Also, the Saudis perceived themselves to be the natural leaders of the Muslim world, or at least the Sunni portion of it, and might have hoped to reduce the ambitions of a potential rival. In any event, the airstrike, coupled with the costs of the war with Iran, ended any Iraqi ambitions to become a nuclear-armed state.

Early in the 21st century, Israeli intelligence agents detected another covert nuclear program in a hostile neighboring state. In this case, the Syrian government had quietly commenced a nuclear program at Deir-es-Zor. They did so by purchasing the assistance of North Korean nuclear scientists and possibly using materials smuggled across the border from Iraq in the run-up to the Western coalition's invasion of 2003. By 2007, the site had clearly developed into a nuclear-processing facility that would be capable of producing weapons for the Syrian government with a few years. Again, the Israeli government determined that such an outcome would be intolerable, particularly given Syrian support for terror organizations in Lebanon that continually targeted Israeli civilians. Also, they once again decided an airstrike would be the only means to destroy the site. However, Syria had learned from its previous dealings with the Israelis and by observing the destruction of Osirak, and, as a result, it maintained a world-class air defense system. The operators of the Syrian IADS remained vigilant for any potential Israeli incursions into their airspace and conducted frequent readiness drills to reassure themselves that they could repel any Israeli attacks.

On the night of September 6, 2007, the Syrian IADS showed no incipient threats, no unanticipated activity in Syrian airspace, and no sign of any hostile Israeli maneuvers. The first hint that anything might be amiss came with the detonation of bombs in and around the nuclear facility. Like

Osirak in 1981, the Syrian nuclear program was destroyed by an IDF airstrike despite the best efforts of Syria's highly trained specialists. How was such a calamity possible, despite all of the Syrian military's precautions? The answer appears to lie with one of Israel's most sophisticated military capabilities, its cyber warfare program. Based upon the reports of Syrian air defense monitors, which did not detect the inbound and outbound Israeli strike aircraft, it appears likely that Israeli agents managed to penetrate the computer systems of the Syrian IADS and insert software that "spoofed" the radar control systems. Essentially, the program informed the Syrian radar systems that nothing had been detected on the radar returns, and thus there was no threat at hand. By the time the Syrians realized that their system had been compromised, it was far too late—the Israeli aircraft had already made it back into their own airspace.

In the immediate aftermath of the reactor attack, the Syrians complained that Israel had committed an unprovoked attack and that the airstrike constituted an act of war. The Israeli government neither confirmed nor denied the allegation, but it did release intelligence photos demonstrating that the site exactly matched a North Korean reactor design. For its part, the Syrian government initially claimed the Israelis had mysteriously bombed an empty stretch of desert and, when reconnaissance photos proved this false, changed the story to claim the site was an unfinished army barracks. Despite Syrian efforts to bulldoze and conceal the site, the International Atomic Energy Agency managed to detect telltale radiation signatures that confirmed the original purpose of the site.

The aircraft used in the 2007 strike were of the same type as those that had Osirak 26 years earlier, although their electronics and armament had been significantly upgraded between the two attacks. Without the major cyberattack, it is unlikely that the Israeli strike force could have penetrated Syrian airspace, flown to the nuclear site, attacked the Deir-es-Zor facility, and emerged unscathed. This operation demonstrated the utility of combining purely machine-based activity (the cyberattack) with traditional kinetic strike assets (the aircraft). It also demonstrated exactly why Israel has spent so much time and national resources to ensure technological superiority over its nearby rivals.

One of the foremost security challenges that has confronted Israel for much of the past four decades has been the ability of armed non-state actors to launch attacks into Israel and then retreat to sanctuaries outside of the country. When external sponsors began supplying advanced missile technology to those groups, the threat became more dangerous. Between 2000 and 2008, the terror groups Hamas and Hezbollah fired more than 12,000 munitions into Israel, primarily from Gaza and southern Lebanon, respectively. Most were fired indiscriminately, aiming for population centers within range. Although the attacks killed only a few dozen Israel citizens, they also did enormous property damage and created a general aura of fear. To combat the threat, Israeli political and military leaders turned to Rafael advanced defense systems, which in 2011 debuted the Iron Dome air defense system. It was the first layer in an envisioned multilayer system to protect against missiles and armed unmanned vehicles. Less than two weeks after it was deployed, an Iron Dome battery intercepted a missile fired from Gaza, and, over the next year, it successfully intercepted 90 percent of its targets.

Iron Dome consists of ground-based radar stations that are wirelessly linked to batteries of interceptor missiles. When one of the radar sites detects an inbound projectile, it tracks the flight path to determine whether it is headed toward a populated area or if it will land in the countryside. If the missile represents a threat, the radar communicates wirelessly to an interceptor battery, which then fires an extremely agile missile capable of bringing down the inbound missile through either a direct strike or a proximity detonation. While the system is not flawless, it has operated well above projections and provided a significant degree of security to Israeli citizens. A directed-energy version of the system, Iron Beam, is currently in development and is envisioned to be in the field by the end of 2018. It will rely upon the same radar systems, but, instead of interceptor missiles, multiple ground-based lasers will fire to converge upon a single target and destroy it.[9]

Israeli unmanned platforms have been in development for the past four decades, and Israel has fielded some very sophisticated systems as a result.[10] Hermes 450 aircraft were sent on a bombing raid against a convoy in the Sudan carrying rockets to Gaza and managed to approach undetected, destroy their targets, and return home without placing a single pilot at risk of being shot down. The largest unmanned platform in the current Israeli arsenal is the TP Eitan, built by Israel Aircraft Industries.[11] It is capable of fully autonomous flight and strike missions and can carry up to 1,000 kilograms distributed upon hardpoints on the wings and fuselage. This aircraft is even capable of ballistic missile interception under certain conditions, making it a very versatile platform. Israel has also experimented with systems largely ignored by the United States, including sensor-laden aerostat balloons and autonomous casualty evacuation aircraft.[12]

THE CURIOUS CASE OF IRANIAN NUCLEAR AMBITIONS

When news of an expanding Iranian nuclear program reached the world's leaders shortly after the Syrian strike, a new norm for attacking a single-point, high-value target had emerged. Of course, the Iranians almost certainly took steps to ensure they could not be victimized in the same manner as the Syrians. When coupled with the distance of the

Iranian nuclear facilities from Israeli airbases and the unwillingness of the United States to allow any Israeli overflights of Iraqi airspace, the Iranians undoubtedly felt much safer than the Syrians had been. Like Syria, Iran possesses a first-rate IADS, one quite capable of detecting, engaging, and destroying Israeli strike aircraft. Iran also has a substantial air force, one that is not as technically skilled as the IDF, perhaps, but which still would operate with the advantage of nearby basing and numerical superiority. There was almost no possibility of repeating the Deir-es-Zor strike against the alerted Iranian facilities at Natanz. Those facilities had significant air and ground defenses, were located in the heart of Iran and thus far from any incursion point, and had been buried deeply underground as an additional physical means of defense against an airstrike. Yet, Israeli policy required some effort to eliminate the possibility of an Iranian nuclear weapon. The Iranian government has repeatedly sworn to eradicate Israel and has invested substantial resources to support Hezbollah and other terror organizations that target the Jewish state. Iran possesses ballistic missiles capable of striking anywhere in Israel; should they be wedded to nuclear warheads, the Iranian government would present a continual mortal threat to Israel.

If a manned strike against the Iranian nuclear facilities could not be launched with anything approaching a guarantee of success, any state hoping to eliminate the Iranian nuclear threat needed to examine alternate means. One method might be fomenting internal unrest in the hope of toppling the theocracy that has ruled Iran since the 1979 revolution. Another method might be to launch ballistic missiles against the facility, although doing so would be an undeniable act of war and hence a violation of the UN Charter. Submarine-launched cruise missiles might be capable of evading Iranian defenses and even be launched with a certain degree of stealth, but their angle of attack and payload capacity would make penetrating a deeply buried site and inflicting catastrophic damage a dicey proposition. Also, even if the Israelis could launch such an attack and maintain plausible deniability, they would almost certainly be blamed by the Iranians and possibly other states in the region and might be subjected to severe retaliatory attacks. How could an attack be launched with complete anonymity and still cripple the Iranian nuclear program? Operational planners began looking for a method to attack Natanz without the Iranians even knowing that they were under attack. To do so, these planners turned toward the machine realm of cyberspace.

To facilitate a successful cyberattack, intelligence operations in nations hostile to the Iranian regime began collecting as much information as possible about the mechanics of the Iranian nuclear program. They found that much like the Syrians, the Iranians had relied heavily upon external expertise to speed the development of their weapons program. Because the nuclear effort was not a home-grown operation, the Iranian scientists and engineers did not have the benefit of the substantial experience that comes from an incremental learning process. As a result, they did not have an inherent and intimate familiarity with the machines they employed on a daily basis, one that might enable them to quickly recognize subtle differences in the behavior of those machines.

Knowing that their nuclear program would be the subject of an intense intelligence collection effort by hostile nations, the Iranians ensured that their nuclear program operated under very strict security protocols that governed both physical and electronic activities. They could not hide the purchase of certain components needed for the program that could not be domestically produced, although they did attempt to hide the purpose behind many of their imports by purchasing dual-use technology rather than items that would be useful only to a nuclear program. To prevent cyber espionage from penetrating the program, Iranian computer security leaders instituted an air-gap system at Natanz. This meant that none of the computers at Natanz was allowed to be connected to the Internet or to any other computers outside of the closed network within Natanz. Again, though, because the Iranian computer network relied upon imported devices and systems, its users had only a basic understanding of cyber security and did not know how to eliminate some of their key vulnerabilities. Many of these weaknesses were inherent to the utilization of commercial, off-the-shelf (COTS) technology, meaning that the Iranian facilities did not rely upon custom-designed computers or unique programs: they used computers that could be purchased anywhere in the world, from major manufacturers, with standard components, connections, and pre-loaded software.

The uranium enrichment process at Natanz relied upon the continual operation of tens of thousands of industrial centrifuge tubes, spun by enormous motors at a very specific frequency. The centrifuges were produced by Siemens Corporation and used a single design, including a programmable logic controller (PLC) that allowed a computer to automate the functions of the motors. The PLCs ran a standard program, Siemens Step7, on the Microsoft Windows operating system. This proved convenient, as it was readily available and familiar to most computer users, but it also meant that those computers needed to be updated regularly with the security patches that Microsoft releases for all registered copies of its software. Because the network was an air-gapped, closed system, it is entirely likely that the Iranians failed to keep it properly updated and hence protected from cyberattacks. It is even possible that the Iranians used illegal copies of Microsoft Windows, not realizing that those copies cannot be updated to remain secure.[13] As almost any computer owner can attest, Microsoft Windows has always suffered from an immense number of vulnerabilities, despite the company's best efforts to find and repair errors in the code. Hackers constantly seek previously unknown vulnerabilities that can be exploited for malicious purposes. When new vulnerabilities are found, they might be directly exploited by the finder or the knowledge of them might be sold to someone else (including Microsoft, which pays a bounty to hackers who find and reveal vulnerabilities to the company). Such new vulnerabilities are called "zero-day exploits," meaning that they were in place from the creation of the operating system and just waiting to be discovered. Other vulnerabilities are occasionally created by the interaction of different programs, but those openings require that the target be operating the correct versions of the software in question to be vulnerable to exploitation.

Hackers are not the only computer users who search for zero-day exploits—national cyber agencies do the same thing, for largely the same purposes. However, rather than seeking information that might be used for financial gain or entertainment, those agencies tend to be seeking the most sensitive of state secrets, things that might seriously compromise state security if revealed to a hostile nation. Finding exploitable gaps in software code enables the likelihood of a successful cyberattack—but they are not expended lightly. Once a new vulnerability is exposed, it will almost certainly be patched in short order, eliminating the opportunity for follow-on cyberattacks using the same methodology, so long as potential victims keep their software continually updated. Of course, if you don't exploit a vulnerability as soon as you discover it, it is always possible that Microsoft will find and patch it, or some other malefactor will use the same opening first. Deciding on when and how to expend a one-time vulnerability can become an extremely frustrating exercise thanks to all of the factors that might influence the decision.

Returning to the Iranian nuclear facility at Natanz, the operators noticed an abnormally high failure rate of centrifuges in 2009. For some reason, the automated PLCs were not spinning the centrifuges at a consistent velocity—instead, they were fluctuating in frequency in such a way as to create destabilizing vibrations that caused the motors to fail. Because the centrifuge operators had only a passing familiarity with the machines, they did not notice that the pitch of the humming machines was changing—a sign of trouble that would likely have alerted many Western workers to something amiss. The effect could be compared to noticing a change in the sound of a car's engine—a driver does not need to be a mechanic to know that *something* is wrong with the sound, even if the driver has no idea what the actual problem might be. Anyone who has spent time at an auto repair shop trying to replicate the sound when explaining his or her fears of a malfunction is likely familiar with this phenomenon. But if a person had never driven a car before, he or she would likely not recognize changes in pitch to be anything outside of the norm—the lack of experience would render the person unlikely to question the sounds heard, even if he or she noticed changes in them.

Somehow, a malicious piece of code was inserted into the Natanz cyber network. There are different theories about how this was accomplished, ranging from a human agent, either a spy or a traitor, deliberately uploading the code to someone inserting a thumb drive found in the parking lot of the facility into a USB port to determine its owner.[14] Regardless of the mechanism, once the code hit the Natanz network, there was nothing to stop its rapid spread throughout the system, where it relentlessly pursued its objectives. The first objective was simply to spread to as many machines as possible, because, like any organic virus, the first goal of a digital virus is to propagate the species, so to speak. The second objective was to check each computer for Siemens-based supervisory control and data acquisition (SCADA) systems. This allowed the virus to target only machines possessing the correct SCADA and PLCs, which indicated they were associated with nuclear centrifuges. When the virus found the correct machines, it rewrote a small section of code to introduce an instability into the centrifuges, making them speed up and slow down their frequencies and quietly damage the entire enrichment program in the process.

Although the Iranians realized something was wrong with their centrifuges, they did not know that the source was a cyberattack. Of course, given that the nuclear weapons program was both illegal and vigorously denied by the Iranian government, the Natanz operators could hardly call in technicians from Siemens to troubleshoot the centrifuges. Not only did the mechanical process begin to fail, the Iranian scientists and engineers began to question their own basic competence—after all, they were failing to accomplish something that the United States and Soviet Union had both achieved in the 1940s—how could they struggle so much on the same tasks several decades later? Not until June 2010, when an obscure Belorussian cyber security company announced it had discovered a previously unknown computer virus dubbed "Stuxnet," did the Iranians begin to realize that they had been attacked through their computer systems. As computer experts began to untangle the Stuxnet code, they determined that it was almost certainly the work of a nation with a very advanced cyber program. Stuxnet relied upon four different zero-day exploits—an expenditure of cyber resources never previously seen in a single piece of malware. Its creators also took meticulous pains to avoid collateral damage, as it targeted only the extremely specific parameters used by the computers at Natanz. And finally, Stuxnet was programmed to delete itself on a predetermined date three years after its original release—plenty of time for it to achieve its ends and possibly disappear if it had avoided detection to that date. Given that it did no damage to any systems that did not fit the parameters of the target, it is highly suspect that the security firm in Byelorussia simply stumbled upon the code—it is far more likely that its creators wanted Stuxnet to be found, if only to send the Iranians hunting after further cyberattacks and trying to figure out who might have brought it into the closed computer network.

The candidates for which nation had both the means and the motive to launch Stuxnet are extremely few, although no nation has formally claimed credit for it. Cyber experts are divided upon the question of whether Israel has the capacity to design such a program without outside assistance. The United States has the resources for such a cyberattack, certainly, as do Russia and China, but none has admitted culpability, probably out of a desire to avoid reprisals from Iran that are unlikely to be confined to the cyber domain. At the same time that Stuxnet came to the attention of Iranian authorities, several leading Iranian nuclear scientists were assassinated, a sure sign that foreign intelligence operatives had decided to accelerate the damage to the Iranian nuclear program. One scholar paraphrased Carl von Clausewitz, when he noted, "Stuxnet potentially represents a case where a computer attack was an instrument of 'policy by other means.'"[15]

Stuxnet raises a number of interesting issues regarding the future of warfare. First, as a piece of software designed to destroy a very specific type of machine, Stuxnet was effectively a digital precision weapon. It delivered the same result as a perfectly placed missile might have, in that it destroyed centrifuges. However, the key distinction is that it garnered the desired effect without killing a single human or even causing any significant collateral damage. As such, there is still a debate over whether an action like creating and unleashing a program similar to Stuxnet's should even be considered an act of violence, much less an overt act of war. Second, the origins of Stuxnet remain unconfirmed, as no nation has stepped forward to take credit. Even if the Iranians wished to retaliate, in the cyber domain or the physical world, they would be unsure of whom to strike. Of course, given Iran's long-documented history of supporting terror groups targeting the West, and Israel in particular, it should be obvious that precise attribution does not particularly matter to the Iranian government. Third, at least in theory, actions like Stuxnet might represent a positive development in warfare. If the damages between conflicting states can be limited to the destruction of machines, no matter how expensive, warfare in the 21st century might be characterized by minimal loss of life.[16]

Although the U.S. government tends to consider Iran a global pariah, that is not always the case. The Iranians are active in global commerce and often cooperate with cyber security initiatives, meaning that when they identify new malware, they can usually be counted upon to alert others, such as the largest Internet security companies, of the new danger. In 2012, Iranian cyber experts, no doubt alerted by their experience with Stuxnet, began reporting the discovery of an enormous piece of malware, soon dubbed "Flame." This newly discovered program had many of the same programming idiosyncrasies, or "signatures," as Stuxnet, suggesting that it might have been produced by the same organization, although Stuxnet was a much smaller and more easily hidden program that had a very precise target, while Flame required 20 megabytes of storage. This size made the Flame file much too large for the traditional means of propagation. Almost every email program has a filter that prevents the transfer of such large files, lest they hog bandwidth and fill servers when a single user decides to forward them to every user on a network. Given that Flame could not simply send itself to new host computers, it needed the active participation of human users before it could reach new sites. Here is where the brilliance and danger of Flame's creators was revealed: the program was disguised as a legitimate Windows security update. It was voluntarily loaded onto target computers by the security professionals tasked with preventing such intrusions.

Once Flame activated itself on a host computer, it proved to be one of the most comprehensive cyber espionage tools in history. Flame included a keystroke logger that could record everything input into the host computer and then transmit that information to a remote server for collection by its creators. If the host machine was not connected to the Internet, Flame could log files of the information it gathered and bide its time, waiting for an opportunity to spill its secrets back to its masters. Flame could also activate the host computer's microphone, essentially turning every infected computer into a perfectly placed bug that could record every conversation in the vicinity. Perhaps the most effective tool in Flame's kit was the ability to activate a computer's camera, allowing the creation of a physical map of all computers infected with Flame—a perfect tool if one had designs on infiltrating or attacking a facility containing Flame-infected computers. When the Iranians announced the discovery of Flame, they did not realize how vulnerable the malware had made their systems. Computer forensic analysis soon revealed that the code had been active since 2006, making it possible that Flame had quietly infiltrated Iranian computers for six years, gathering data, eavesdropping on conversations, and surveilling anything within reach of a computer's integrated camera (another feature of COTS computers that probably has no place in a nuclear facility). Not only did this give Iranian intelligence operatives fits as they tried to determine what information might have been collected, it also put them in a serious cyber quandary. If computer users must routinely update their operating systems to protect themselves from malware but it is possible that that security update is itself the source of the worst infection of all, what should the cyber professional do to secure a network? Risk continual penetrations to avoid falling victim to an even worse piece of software? Operate without computers in any sensitive locations? Or attempt to develop a domestic computer operating system without the help of any outside experts in order to stop using Microsoft Windows? Thus far, the Iranians have not solved this quandary, as there really is no obvious answer for a nation without the decades of programming experience found in Western countries.

In 2012, Iran may have launched a relatively crude but effective cyberattack in retribution for the Stuxnet and Flame penetrations. Interestingly, though, the target of the attack, the Kingdom of Saudi Arabia, was definitely not responsible for either of the earlier operations. It might have simply been a proxy target, one more accessible to Iranian hackers, or the attack might have been to deter Saudi cooperation with the United States. The attack, dubbed the "Shamoon" virus, targeted computers networked at Aramco, the Saudi national petroleum corporation. It propagated itself as quickly as possible and then, at a predetermined date and time, every copy of Shamoon executed a file-wiping command that simply destroyed data and rendered the computer that held it useless. Almost immediately, 30,000 workstations were rendered worthless. One cyber security expert referred to them as "bricks that could not even be booted up."[17] While Aramco had the financial resources to rebuild its network in a matter of weeks, the corporation still lost tens of millions of dollars from the attack and the resulting production delays. The author of Shamoon has not been positively identified, although a group calling itself Cutting Sword of Justice claimed responsibility for the attack. Several signatures within the program pointed to Iranian origins, but it is possible that these tidbits of information were planted by a third party solely to throw suspicion onto Iran.

Iran has not devoted as many resources to military robotics as it has to its nuclear weapons and cyber programs, but it has still developed a line of unmanned aerial vehicles called the Ababil series. These platforms vary in size and are primarily designed for intelligence, surveillance, and reconnaissance missions, although they can be weaponized if desired. Iran has supplied several variants to Hezbollah, which in turn has used them for scouting and target spotting over Israeli territory. The Israelis have shot down several Iranian-built Ababils, normally in the vicinity of the northern border with Syria and Lebanon.

KOREA: THE WORLD'S MOST DANGEROUS PENINSULA

The idea of deploying a force composed primarily of semiautonomous machines might be exceedingly attractive, particularly in locations that are difficult to reach for reasons of distance or geography. In particular, locations that have seen a long-term U.S. military presence, such as the border of the demilitarized zone (DMZ) of the Korean Peninsula, might be a logical place to deploy such machines. The necessary command and maintenance personnel associated with fielding a robotic force might be sent on a rotational basis, although, in theory, a large amount of their work might be performed from a significant distance.

The South Korean military has used sentry robots as part of their patrols of the DMZ for years and clearly has accepted the presence of military robots as one aspect of perimeter defense. One of their most prominent systems, the SGR-A1 robot, is a fixed-location platform outfitted with advanced sensor systems and firearms. It can be placed in an alert mode, in which it will notify human operators of any detected intrusions, but it can also be given full autonomy to engage any infiltrators working through the DMZ. Naturally, the enhanced sensor capabilities of automated sentries, as well as their ability to communicate with one another and other sensor platforms, also lends credence to the concept of replacing human forces with machines that will not become bored by the tedium of remaining in constant vigilance. Because machines are inherently less valuable than human lives, sentry robots along the DMZ could be limited to nonlethal models, armed at most with taser devices or incapacitating chemical agents. Of course, given the history of relations along the DMZ, such restraint might rest upon a fragile foundation, and thus armed variants could quickly appear during a crisis. South Korea has relied for years upon unmanned aircraft to patrol the extensive coastlines and the DMZ region, constantly seeking to detect any aggression from its northern counterpart on the peninsula.

North Korea seems determined to master the tenets of 20th-century warfare rather than attempting to match the technology of its rival. Between active duty and reserve forces, the North Korean military incorporates fully one-third of the North Korean population. A relentless drive toward the production of nuclear weapons has led to at least a handful of successful atomic detonations, and, in 2017, North Korean president Kim Jong-Un claimed a capability to place nuclear warheads upon newly developed intercontinental ballistic missiles. This claim triggered a heated exchange of inflammatory rhetoric with U.S. president Donald Trump, leading to increased tensions throughout the region. While it is impossible to definitively state the status of North Korean robotics developments, the isolation of the country and its relatively weak industrial, economic, and scientific base make it unlikely that its military robots will be much of a match for any of its likely competitors.

CHAPTER 8

The Road Map

To inquire if and where we made mistakes is not to apologize. War is replete with mistakes because it is full of improvisations. In war we are always doing something for the first time. It would be a miracle if what we improvised under the stress of war should be perfect.

—Hyman G. Rickover, 1964

Making predictions about the future of warfare is a dangerous game. As Secretary of Defense Robert Gates noted in 2011, "When it comes to predicting the nature and location of our next military engagements, since Vietnam, our record has been perfect. We have never once gotten it right, from the Mayaguez to Grenada, Panama, Somalia, the Balkans, Haiti, Kuwait, Iraq, and more—we had no idea a year before any of these missions that we would be so engaged."[1] Predicting the direction in which technology will develop or how it will be used in future conflicts is an equally fruitless activity. However, creators of fiction have spent a substantial amount of time and energy fantasizing about the future of military robotics and, in many ways, have shaped our expectations for the future as a result.[2]

LITERARY SCIENCE FICTION AND THE FUTURE

In point of fact, the term "robot" came from the stage, coined by Czech playwright Karel Čapek in his 1921 play *R.U.R.: Rossum's Universal Robots.* The term, which came from the Czech word "robota," meaning a serf's tedious manual labor, referred to Čapek's vision of a future society dominated by artificial life forms first created as personal servants. These self-aware machines, not content with their existence as machine slaves, launched a rebellion, annihilated humanity, and created a new, mechanical society.[3] This dystopian future, in which humans are overthrown as the dominant force on the planet by their robotic creations, has remained a major theme in science fiction for nearly a century.[4]

Science fiction authors have devoted an enormous amount of their production to considering the future applications of military robotics. While thus far none of them has proven to be precisely correct, many have offered broad concepts that bear a certain amount of examination, if only to illustrate some of the unexpected consequences of outsourcing war fighting to military robots. At one extreme, the steadfastly optimistic Isaac Asimov postulated that virtually all negative consequences of a robotic revolution could be avoided with three overriding program commands integrated into all robots. At the opposite end of the spectrum, Fred Saberhagen envisioned a universe in which humans struggled for survival against killer machines run amok. Somewhere in the middle of these extremes, an interested reader will find Douglas Adams, who used robots primarily for comedic relief in his most well-known works, and Jack L. Chalker, who outlined a future in which machines seized control of human civilization in order to prevent mankind from destroying itself. Each of these four authors sold millions of copies of their imagined futures, and all of them offer both cautionary tales and reasons for hope as humanity hurtles toward a greater integration of robotics into everyday life.

Isaac Asimov's "three laws of robotics" set the basic parameters for robots in his *Foundation Trilogy*. Perhaps, the classical science fiction writer offers the simplest guidance for this issue in his 1942 short story "Runaround," though he had hinted at these concepts in earlier fiction. In Asimov's envisioned future world, all robots are governed by three simple programming rules that are hardwired into their systems and which offer a hierarchical set of basic controls for all robots, regardless of the level of their autonomy or the intent of their function. These laws were very straightforward:

1. A robot may not injure a human being or, through inaction, allow a human being to come to harm.
2. A robot must obey the orders given it by human beings, except where such orders would conflict with the First Law.
3. A robot must protect its own existence as long as such protection does not conflict with the First or Second Laws.[5]

Of course, some potential loopholes can still be found, a matter that Asimov considered over the course of dozens of books on the subject. For example, the first law creates the possibility that a robot may be paralyzed to inaction if it is faced with a situation in which it can only save a limited number of a larger pool of victims. The second law essentially makes all robots communal property, a highly unlikely outcome in a capitalist consumer society. Later, Asimov wrote an addendum to the three laws, which he dubbed the "zeroth law of robotics" to maintain the order of precedence:

0. A robot may not injure humanity or, by inaction, allow humanity to come to harm.[6]

This prime directive recognized that individual humans might have to endure harm for the greater good of society as a whole, although a robot might define an injury to humanity in a far different fashion than how humans might choose to do so.

Of course, the three laws are nice in principle, but in practice they would seem to preclude the use of robotic weapons. A weapon that could not under any circumstances cause harm to a human being would have no practical utility on the battlefield, as it could be circumvented through the use of human shields. Likewise, a military robot that must follow the orders delivered by human beings could simply be told to shut down by any enemy capable of communicating with the device. Finally, a military robot that could rarely if ever place itself in any danger would hardly serve to remove the danger to humans on the battlefield—a major part of the attraction of military robots is that they might perform the most dangerous tasks, the ones that would be unacceptable to order a human being to perform. As a result, at least in the field of military robotics, Asimov's laws, or anything resembling them, will remain in the realm of fiction.

Saberhagen's *Berserker* series, first appearing in 1967 and eventually including 17 volumes, was set in a future universe in which humanity has spread out to the stars, establishing a human empire spanning dozens of worlds. As part of that expansion, humans discovered previously inhabited worlds that had been scoured of all life by an unknown but obviously powerful entity. Further investigation showed that two alien civilizations had engaged in an interstellar war of racial survival. In a last-gasp effort, one side created and deployed the berserkers, killer machines ordered to eradicate all enemy life. Unfortunately, through deliberate sabotage, faulty programming, or copying enemy technology, the berserkers not only annihilated the enemy but then rededicated themselves to the elimination of all life in any form, wherever it could be found. Humans simply blundered into space occupied by berserkers and suddenly found themselves in a no-holds-barred war against extinction.

Saberhagen's series is worthy of consideration because he investigated the "law of unintended consequences." The alien creators of the berserkers, long since destroyed by their own creations, became so dedicated to the quest for annihilating their immediate enemies that they gave little thought to the possibilities of creating a series of killing machines and releasing them on an unrestrained basis. They apparently failed to place limits upon their creations that might prevent out-of-control self-replication, and they either included no failsafe self-destruct systems or they triggered them before it was too late. Saberhagen's series permutated into an almost limitless series of tangents as he envisioned as many mechanisms of conflict as possible. Having devised an entire alternate universe, he chose not to end the series, leaving open the possibility that other writers might continue operating in his crafted environment.

Douglas Adams is best known for his science fiction comedy series *The Hitchhiker's Guide to the Galaxy*. In this series, robots are simply one more element of the cast of characters, with Marvin the Paranoid Android playing a central role. Marvin's comedic function largely revolves around his designer's decision to provide emotions to his robots, a function that caused Marvin to be prone to fits of chronic depression. While the inventors of these emotional machines believed it would make their robots more "relatable" for humans and hence more marketable and profitable, they were surprised to discover that emotionally self-aware robots might reflect upon their own situation with extreme disappointment. Marvin frequently reminds his companions that he is far smarter than they are and thus his role in life, serving essentially as a robotic butler, is less than fulfilling. He knew that he could solve many of the fundamental questions of the universe; but he was being asked to fetch beverages and chauffer guests, which caused him to fall into a deep reflection. He becomes a source of frustration for the protagonists at times, as they must pay him compliments and beg for his assistance before he can be cajoled into emerging from his internal doldrums and offering any assistance in solving the problem at hand.

Adams's discussion of future robotics is useful because it reminds us that attempting to replicate human behaviors and thought processes in machines is likely to result in frustration for everyone involved. While some experimental robots have been designed to replicate emotional responses and to recognize emotional cues in humans, they cannot be said to truly feel the emotions they display. Yet, it might not be possible to replicate human thought patterns without the incorporation of emotions into the mix. There are fundamental differences in the strengths and weaknesses of human and robot cognition, and any attempt to imbue robots with all of the attributes of their creators is likely to simply introduce new avenues for failure and unexpected negative behaviors. Within Adams's prose, purpose-built machines show none of the failings evidenced by Marvin; they simply go about their functions and remain largely in the background, enablers rather than colleagues, eliciting neither sympathy nor irritation.

Jack L. Chalker's *Rings of the Master* series is set in an unknown number of centuries, or even millennia, into the future. In his imaginary universe, a team of international scientists, operating in the near future, built a massive supercomputer designed to oversee all of the defense functions of the United States and its allies. Dubbed "Master System" by its builders, the computer had the primary task of preventing the eradication of humanity. In practice, its laudable goal was to protect people from the ravages of warfare. While it might have control over military systems, it could use them only for defensive purposes in pursuit of its ultimate aim, saving humanity by ending war. The programmers included a failsafe mechanism—any five users could plug override devices into the master control board and gain control over Master System. One fateful morning, they turned it on—and it immediately set about fulfilling its purpose. However, Master System had come to a radically different conclusion than its creators about how to save humanity. First, it seized control of the military systems around the globe and eliminated any potential threat to its own systems, a step that included killing its creators. Next, it instituted a forced human exodus to the stars, on the assumption that the preservation of the human race would be far easier if they were living on thousands of planets. To assist in the transition, Master System genetically altered the colonists to suit the ecology of their new home worlds, stretching any conventional definition of "human." Finally, it sought to eradicate almost all advanced technology, particularly upon earth, guaranteeing that humans would not be able to rise to a level at which they might represent a threat to themselves at any point in the future. Master System maintained control over interstellar trade, military forces, and civilizational advance. Chalker's main protagonist became accidentally aware of the full extent of Master System's activities, a forbidden knowledge that made him immediately the target of highly specialized assassination robots. He fled and in the process assembled a team of insurgents to seek out the five override devices. Master System's programming required these to be maintained in human hands, but it hid the nature of the devices and spread them throughout the galaxy. Over the four books of the series, the protagonists visited dozens of worlds searching and in the process learned more of Master System's activities on behalf of the goal to preserve humanity.

Rings of the Master reminded its readers that it is impossible to plan for every potential outcome, a situation that is particularly relevant when considering computer programming. Just as the creators of Master System failed to anticipate that it would take its mandate so literally or that it would execute its programming so ruthlessly, today's programmers of military robots are unlikely to foresee every situation their creations might face. Even if a program is executed flawlessly, it will still include gaps created by unexpected circumstances. While humans often fail when confronted by uncertainty, they have a lifetime of experience to draw upon in their decision making. Also, we have come to expect and accept a certain degree of failure from fellow humans and account for it in our planning. When it comes to machines, though, we have become increasingly expectant of a flawless performance and anything less produces frustration, anger, and reduced faith in the system being evaluated.

VISUAL MEDIUMS AND SHARED EXPERIENCES

Books are not the only mechanism by which science fiction concepts enter the everyday consciousness of society, of course. In particular, movies and television programs reach millions of consumers, offering a shared experience that can shape our expectations of the future and our potential interactions with autonomous machines. These formats offer the opportunity for exciting visual representations that exert an outsized influence on viewers' perceptions of technology. However, by definition, a movie, with its relatively limited run time, cannot examine a subject with the richness of detail contained within a novel, nor does it engage the imagination of the consumer in the same fashion. If a novelist describes a character, an action, or an object, that description forces the reader to imagine it, and that imaginary vision will be shaped by one's own perceptions, preferences, and experience. One of the highest-selling authors of the 21st century, Stephenie Meyer, deliberately avoided supplying a detailed description of her protagonist, Bella. This allowed readers, most of them young females, to envision themselves at the heart of the story, without being contradicted by the author, who described the physical attributes of every other character in excruciating detail. Not until the books were converted into movies did the reader discover the definitive appearance of the main character, and many disagreed with the casting choice when it did not match their imagined versions. Those disagreements did not stop the five movies from grossing more than US$3 billion worldwide, making it one of the most lucrative film franchises in history.

There have been a number of science fiction movies and programs that have exerted an enormous influence over the public perception of robotics and artificial intelligence. They range from the lovable and benevolent robots of the *Star Trek* and *Star Wars* franchises to the apocalyptic villains of the *Terminator* and *Battlestar Galactica*. Each of these popular offerings has exerted an enormous influence over multiple generations, and their ability to continue producing lucrative new content demonstrates that they have tapped into an element with enormous popular appeal.

In *Star Trek*, the franchise competing most closely with *Star Wars* for the loyalty of fans worldwide, robots and artificial intelligence play a radically different role. In the first iteration of *Star Trek*, airing from 1966 to 1969, computers played a background role, particularly the part of the ship's computer. Despite being set three centuries in the future, computers apparently had not advanced much, as many of the depictions of them involved vacuum tubes, flashing lights, and enormous amounts of space needed to support relatively basic functions. Given that the first working integrated circuit was demonstrated in 1958 and that Moore's law was postulated in 1965, the evidence was already available that computers would develop far faster than the show's creators realized. The original show's computers seem silly to modern viewers because they were so unnecessarily primitive even a few years after the show went off the air. Even in the rebooted franchise *Star Trek: The Next Generation*, airing from 1987 to 1994, the predictions of computer advancement and artificial intelligence proved hopelessly far behind the developments of even the past two decades. Yet, the series is noteworthy to the current study for a few reasons. First, many of the technologies taken for granted by the actors of the series did not exist at the time of filming but have become ubiquitous in modern society. For example, tablet computers made their first popular-culture appearance on board the *USS Enterprise*, the setting of the show, but are now the fastest-selling form of personal computer. Replicator machines, which instantly produced everything from food products to spare parts, have now begun to emerge in the form of 3-d printers. Holographic entertainment systems, a major feature of the show, are currently in development. Second, an integral character of the series, Lieutenant Commander Data, is a robot designed by a reclusive genius who spends the majority of the series attempting to become more human, both in appearance and behavior. In the process, he highlights many of the social constructions of what it means to be human and triggers comparisons of machine and organic capabilities. Often, the lesson of an episode revolves around the inherent shortcomings of humans and the predicable behavior of machines.

In the *Star Wars* series of films, only one character has appeared on-screen in all eight films to date, the protocol robot C3P-O. Although often referred to as "he" by fellow characters, C3P-O is an android and hence has no gender. It repeatedly reminds the viewer that it is fluent in over 30 million forms of communication, although its translation services are rarely needed in a universe where almost every species seems able to understand one another, except when the situation absolutely calls for confusion or obfuscation. C3P-O is accompanied by a trusty sidekick, R2-D2, a two-legged cylindrical robot that communicates via an array of beeps and grunts. The two become intimately associated with a burgeoning rebellion against a galaxy-spanning empire and discover an unlikely penchant for heroic acts despite essentially being specialized service robots. Merchandise associated with the films has always sold extremely well, including items featuring the two robots. In 2007, the U.S. Postal Service used R2-D2 as the centerpiece of a publicity campaign, largely by repainting mail drop boxes to resemble the beloved robot. *Star Wars* showed a universe in which humans and robots interacted seamlessly, with the humans in firm control over their machine companions in almost every situation.

On the robotic apocalypse side of the entertainment ledger, *The Terminator* is undoubtedly the most successful such movie franchise, grossing nearly US$2 billion worldwide. What began as a fairly low-budget science fiction film with the first installment quickly grew into a major box office blockbuster series, with the movies known for their trademark cutting-edge special effects and their time-traveling hunter-killer cyborgs. The premise of the movies, that a future robot revolt against humanity would conclude with a struggle for basic survival, resonated with audiences living on the cusp of the robotics revolution—and the perfect human appearance coupled with cold machine logic struck a chord with many viewers.

The television program *Battlestar Galactica*, first running from 1978 to 1979 and later rebooted from 2004 to 2009, relied upon a robot uprising against human control to establish the parameters of the series. Self-replicating robots, called Cylons, commenced an apocalyptic first strike against human military systems in an effort to eradicate humanity. In both versions of the show, a surviving spaceship, the *Battlestar Galactica*, survived the attack and commenced fighting against the genocidal robots. In the show, anything exhibiting artificial intelligence was to be feared—the namesake spaceship survived only because it contained obsolete technology and was thus ignored by the initial attack. For dozens of episodes, the creators of the show examined the struggle between human ingenuity and machine efficiency. Given that the target audience consisted entirely of humans (so far), it is unsurprising that the protagonists eventually found not only a means to survive but to eventually triumph. Along the way, the show examined themes of conquest and collaboration, espionage and sabotage, and the perils of closely duplicating one's enemy in the quest for eventual victory. Like the other franchises, it also shaped certain elements of technological development—for example, the U.S. Air Force flies the F-16 "Fighting Falcon," which has been affectionately nicknamed the Viper by its pilots for its resemblance to the space fighter craft used in the *Battlestar Galactica* universe.

WARS OF ANNIHILATION

The technology needed for a war of annihilation is not far-fetched. Rather, with a few modifications of existing machines, it already exists. In fact, this annihilation concept is far easier to fathom than the much-greater difficulty of engaging in a discriminatory form of conflict. Combining a few existing robotic systems, it is not difficult to imagine an army of annihilators. One might begin with the centibots search-and-observation platform. The centibots project sought to create a very simple swarm of ground robots that could quickly explore and map very large and complex areas. Each machine was relatively simple and unsophisticated, as was its programming. Thus, each centibot component required few resources. Its commands were to continually explore the local environment, to maintain distance from other members of the swarm, and to report its observations back to a central processor. With these simple rules, a group of 100 centi-bots can map every structure in a medium-sized city in only a few days. A slight change in the programming might be to report the presence of any detected humans, essentially serving as a beacon for on-call killing machines. Alternately, the centibots program might be added to an armed robot design, allowing each to serve as an independent hunter-killer, a tireless searcher whose only function is to eliminate the enemy. A third variant would be to incorporate an explosive into each centibot, to be det-onated in a suicidal fashion when humans are encountered. This approach might prove far more costly in resources, but it would also be simple to implement with existing technology.[7]

Given the behavior of members of the Islamic State (IS) in occupied ter-ritory, it requires little imagination to foresee an increase in urban combat in the future. Urban warfare largely negates the inherent advantages pro-vided by U.S. military technology, providing enemy forces a much greater opportunity to inflict large numbers of casualties upon American forces. When coupled with the American public's well-documented reticence to accept a large number of U.S. casualties, it is easy to envision a scenario in which combat troops facing an entrenched enemy are replaced by combat robots. However, the speed of modern warfare, coupled with the challenges of maintaining command and control over remotely operated vehicles, all but guarantees that such combat robots would need to operate with a substantial degree of autonomy if they are to be effective in the fight.

Taken to its logical extreme, the utilization of ground combat robots operated in large numbers presents a terrifying vision of the future of war-fare. At the same time, American political leaders are increasingly hesitant to commit to large deployments of U.S. troops to the Middle East or other danger zones in the world, particularly in the aftermath of the invasions and occupations of Afghanistan and Iraq. Yet, victory by an organization such as IS, which claims it plans to establish a global caliphate capable of competing with any other major power, presents an intolerable situa-tion. Airstrikes served to slow but not stop the advance of IS forces across Iraq and Syria, and the military forces of those nations initially proved incapable of standing up to IS attacks. IS has proven particularly effective at spreading terror in advance of their marches, releasing footage of hor-rific executions of captured troops and civilians for a variety of offenses. Rather than stiffening resistance to the IS movement, these propaganda pieces have managed to recruit thousands of foreign fighters to the IS ban-ner and induce the desertion of tens of thousands of Iraqi troops. A coor-dinated assault of Iraqi ground troops, backed by coalition airpower and with advisors on the ground, gradually cleared the IS stronghold Mosul but took enormous casualties in the process.[8]

The behavior of IS occupying forces leaves little doubt that the organi-zation considers its struggle a fight to the death. For example, the decision to burn alive a captured Jordanian fighter pilot, an act that was recorded and distributed via the Internet, demonstrated that IS has no interest in adhering to any of the international laws of armed conflict and that it does not fear provoking the moderate Muslim regimes within the region. In short, IS will stop at nothing to achieve its dreams of conquest. This level of fanaticism presents a special challenge to any nation attempting to combat IS while remaining safe from retaliation. Already, IS operatives and sympathizers have launched dozens of terror attacks around the world, mostly targeting elements of the civilian population. These attacks amply demonstrate that IS has no intention of hindering its own efforts by following a system of rules created by the most powerful states in the world—those that have already reached the position of dominance that IS hopes to reach—as following the rules of warfare would create an almost insurmountable hurdle for its efforts. Given the near-unanimous opposition of the rest of the world, IS essentially has nothing to lose by launching terror attacks or using any other means to coerce the governments of the world into compliance with its objectives. Further, it might be possible for terror organizations like IS to gain access to robotic weapons in the near future, a concept advanced by Lambér Royakkers and Rinie van Est, who note, "The effects of armed military robots that are in the possession of fundamentalists or [terrorists] could well be devastating and would, by comparison, pale the impact of roadside bombs."[9]

Of course, IS has plenty of disadvantages, and a major one is its lack of technological sophistication. Thus far, IS possesses no conventional airpower of its own and only rudimentary antiaircraft weaponry. While IS forces have captured a substantial amount of heavy weapons, including artillery and armored vehicles, they do not possess the means to produce these items or even to resupply their ammunition and the parts required for basic maintenance. Needless to say, without an external nation-state sponsor, IS will be unlikely to develop such capabilities, and it certainly has no ability to design, build, and field military robots. The organization has demonstrated an ability to improvise, though, using commercially available toy drones to fly scouting missions over Iraqi positions, and, on a handful of occasions, to drop grenades upon concentrations of Iraqi troops.[10]

If the IS willingness to torture and kill captured enemies is one of its main tools to inspire terror and continue its advances, perhaps the solution is to eliminate its supply of captives. By employing only unmanned systems against IS, the United States and its coalition partners in the fight against IS can at least guarantee that they will not be forced to stand idly by while captured members of their own forces are maimed and murdered. Thus far, the United States has proven unwilling to send more than a few hundred ground forces into the IS-dominated regions, although it has launched hundreds of manned airstrikes against the group, putting pilots into at least a small amount of danger. The vast majority of

U.S. anti-IS missions have been conducted by unmanned aerial vehicles, though, which have managed to kill hundreds of IS fighters and at least a handful of high-profile IS commanders.

Would a large force of ground robots fare well against IS? Probably not, if they relied upon remote operators to choose when and where to apply force. The small delay caused by such controls, as well as the incomplete perceptions and susceptibility of ground robots to electromagnetic interference or jamming, means the current models would probably not be capable of turning the tide on the ground, although they might serve as force-multipliers in certain situations. On the other hand, they could hardly perform worse than the Iraqi Army did upon first meeting IS forces. Despite holding a significant advantage in both numbers and equipment, the Iraqis collapsed before Mosul and were pushed back to the outskirts of Baghdad before halting the advance. It seems that more than a decade of U.S. and coalition training did little to instill confidence into the Iraqi military, which steadily retreated in the face of IS advances. It is little wonder that IS forces claim divine protection, as they have piled success upon success in their march across Iraq. The capture of Mosul, Iraq's second-largest city and the capital of a major oil-producing region, certainly showed that IS has become a force that cannot be ignored or simply brushed aside and triggered renewed coalition involvement in the region.

THE FUTURE OF ROBOTIC FORCES

There are fundamental limits that govern the size and the behavior of American military forces, but those limits might not apply quite so well to military robots. At the end of the fiscal year 2017, the legal limit upon the size of the U.S. Army will stand at 460,000 troops, but that number only applies to humans in uniform.[11] In the 1990s, the military began turning increasingly to contractors to circumvent this problem, outsourcing many of the mundane aspects of running a military force as a means to reserve more uniformed slots for warfighters and functions that could not be trusted to a contractor. Military convoys, mess halls, and communications largely became the province of contractors, with the number of roles filled by non-uniformed personnel growing on a yearly basis. As automation has swept into the industrial sector of the U.S. economy, it has also become an increasingly important aspect of the military. After all, there are essentially no limits upon the number of military robots that can be incorporated into the armed forces. While the initial costs of procurement might seem prohibitive, those costs are susceptible to economies of scale in ways that human recruitment costs are not, and the latter tend to be largely upfront sunk costs. Once a robot has been purchased and included into a military unit, its maintenance costs tend to be relatively low, whereas human soldiers only get more expensive the longer they are in service.

Robots offer the great solution to the conscription problem faced by Western militaries. In the United States, the armed forces shifted to an all-volunteer model in 1973, though not without howls of protest from the commanders of each service. Congress imposed the shift, a move that required military organizations to spend a great deal more time, effort, and resources on the recruitment and retention of personnel. This led to a significant rise in the pay rates of military members, an increase in the available benefits, and a greater emphasis upon professional troops serving for an entire career rather than short-term service members, whether conscripted or voluntary, who then entered into other professions. There has been a corresponding shift in the makeup of the officer and enlisted corps, particularly regarding the socioeconomic demographics of each group. There has also been a massive decline in the number of veterans within the political class and in the number of top elected officials with a child serving in the military. When the United States declared war on Germany, Italy, and Japan in 1941, it was presumed that the entire resources of the nation would be committed to the fight, including the offspring of congressional members. When the United States invaded Iraq in 2003, only a handful of members of Congress had a son or daughter in the military and less than 30 percent of congressional members were veterans.[12]

Military robots can theoretically solve the demographic problem faced by the military services. Given that less than 8 percent of American citizens have chosen to don the uniform of the U.S. armed forces, robots might alleviate the shortfall.[13] They can be mass produced and committed to dangerous or boring military operations with little regard for their ramifications within the force. After all, even if thousands of robots are destroyed in an engagement, there will be no funerals, no grieving parents and widows, no payment of death benefits, and no media pressure to end the war as quickly as possible.[14] Rather, there will simply be an economic cost to replace the destroyed hardware, and the war can rage onward. Unfortunately, this mind-set might also lead to a presumption that wars can be fought with impunity—if there are no human casualties, there is likely to be little outcry. Certainly, images of dead civilians might provoke remorse within the U.S. civilian population—but will it be enough to drive the United States out of a war? Dead civilians in Syria were not enough to get the United States directly involved in that conflict, so why would they have a greater effect solely because they were killed by American-built and operated robots? Medea Benjamin considers it highly unlikely, noting "If the killing of a sixteen-year-old American fails to spark any substantial debate in the US media regarding the blatantly extrajudicial nature of drone attacks, then certainly the killing of poor Yemenis or Somalis is not going to cause a stir."[15]

ROBOTS BECOME KEY MILITARY ASSETS

One interesting aspect of military robots is that they do not suffer from many of the common failings of human troops. Just as industrial robots are utilized because they do not suffer from fatigue or boredom, military robots can be called upon to perform tedious or dangerous tasks with no drop-off in performance over time. Because they are immune to fear, they will not flinch at receiving orders that would appear suicidal for organic forces. Their loyalty, such as it exists, should never come into question, nor should robots present any problems with desertion, an issue that has plagued military units for millennia. Yet, these strong points in favor of military robotics are not without their drawbacks—although they will not voluntarily desert, per se, programming errors might lead to essentially the same effect if robots wander away from their intended mission. Likewise, if military robots are not fully autonomous, they require some degree of external input and control. At least in theory, that control mechanism might be hijacked, causing the robot to accept unanticipated and undesirable commands, including turning upon its unit. At the very least, any control signal might be subject to disruption, jamming, or a loss of connection, a situation that might render the affected robots completely or partly useless.

Military commanders have always been forced to consider the loyalty and morale of the troops under their command and faced enormous peril if they failed to do so or misunderstood the level of commitment to the mission felt by their troops. Even the greatest military commanders in history have suffered from their share of mutinies, desertions, and betrayals, something that their intellectual descendants have extensively noted. Consider the case of Alexander of Macedon, widely considered the greatest of ancient commanders. His armies swept through Greece, the Persian Empire, and southwestern Asia and seemingly no opponent could hope to stand against his forces. Even when outnumbered five to one at the Battle of Gaugamela, Alexander's forces routed those of Darius II of Persia while taking minimal losses. Yet, for all his battlefield prowess, Alexander faced a mutiny at the Hyphasis River in 326 BCE. The event illustrated that tactical victories and even strategic conquests cannot guarantee that troops will remain motivated and eager to stay upon a campaign for an indefinite period. The true mark of Alexander's leadership was his ability to hold his forces together. However, he could not stifle his own quest for glory and embarked upon ever-greater efforts to expand his empire. His exhausted Macedonians, the elite core of his army, demanded the opportunity to return home and enjoy the spoils of war rather than face the dangers of more fighting. This failure of morale, rather than any battlefield defeat, forced Alexander to abandon his plans to conquer all of Asia.[16]

Military robots, however, are immune to complaints or homesickness, which makes them an inherently attractive option, particularly for a protracted campaign. When in 2006 the U.S. Army announced that troops deployed to Iraq would have their tours extended from 12 to 15 months, morale among U.S. forces plummeted. While outright desertion did not appear to result, the decision certainly hurt the troop recruitment and reenlistment numbers, and a number of prominent officers, retired personnel, and defense intellectuals pronounced the U.S. military to be exhausted by the increased tempo of operational deployments and extended tours. When compared with Alexander's troops, many of whom remained on campaign for an uninterrupted period of 13 years, the 21st-century complaints might have seemed ridiculous, the whining of a pampered generation of warriors. Yet, to an all-volunteer force, the expectation of balancing deployments with rotational cycles back in the United States had become the norm, and that system was certainly not created by the troops affected by the change.

Of course, desertion is not the only difficulty that human commanders have faced when attempting to wield military power. Humans have proven quite open to betrayals under certain circumstances, ranging from fear of death to pursuit of fiscal rewards. During the U.S. Civil War, thousands of prisoners of war (POWs) held by each side of the conflict faced years of confinement in wretched conditions. The mortality rate of the worst prisons rose above 25 percent per year. Unsurprisingly, when recruiters began visiting the prison camps and offering releases in exchange for switching sides, they found a receptive audience. Almost 6,000 Confederate POWs agreed to join the Union Army, as long as they would be sent to the Western Frontier rather than being deployed against their former compatriots. This made it less likely that they would return to the Confederates at the first opportunity or that they would be captured and executed by the Confederates for changing sides. It also freed up an equal number of U.S. Army regulars from frontier posts for service against the Confederacy. Numbers for Union POWs who exchanged their blue uniforms for gray are more difficult to determine, in part because Confederate record keeping was substantially worse. However, at least several hundred left Andersonville Prison to take service with the Confederacy, despite the sure knowledge that they would face Union troops in battle. Many claimed that switching sides offered their only hope of survival and that they intended to desert their new units at the first opportunity. Nevertheless, their presence swelled Confederate ranks at a critical point in the war, and all might have been court-martialed for treason or at least desertion had President Lincoln not intervened on their behalf.[17]

Robots might not offer the likelihood of desertion, but they certainly carry a substantial risk of outright betrayal under certain conditions. By definition, computers are designed to accept programming from an external source, and this creates an inherent vulnerability for robots. If a robot's programming can be altered, its behavior will naturally follow suit. Even without actually changing any aspect of a computer program, the very nature and assumptions of the program at hand can create a vulnerability. For example, many remotely piloted aircraft have a failsafe program in the event of losing the control signal. For some models, the aircraft returns to the geographic position in which the signal was last received, a simple concept that increases the opportunity to regain control of the aircraft. Such a failsafe mechanism also makes inadvertent penetration of sovereign airspace far less likely, which is a significant concern in the crowded international airspace of the 21st century. Other failsafe mechanisms might send an out-of-control RPA into an orbit pattern, seeking to resume contact while remaining in a predefined area. The advantage of this approach is that it offers a better chance of resuming communications if the source of signal loss is deliberate interference from the ground—by moving, the aircraft has a better opportunity to avoid the interference zone and resume normal operations. Flying machines with limited autonomy might essentially continue with the mission while attempting to reestablish contact or might be programmed to select from a series of options to defense from enemy interference. This might include taking evasive actions and attempting to reestablish communications, or it might involve seeking the first safe location in which to land and await recovery efforts.

When some of the most advanced systems lose communication links, they can guide themselves to a friendly landing site. Such self-guidance relies primarily upon receiving and correctly interpreting a GPS signal. If the communication problem is due to a physical component breakdown, receiving the GPS signal might be problematic, but assuming there is nothing wrong with the RPA's ability to receive the GPS signal, the aircraft should be able to pilot itself to safety and land. However, the GPS signal is a fairly weak transmission—jamming a GPS signal is easily accomplished. Overpowering the signal with a stronger broadcast on the same frequency, on the other hand, is somewhat more difficult but can pay enormous dividends if done correctly. Such "spoofing" of a signal might convince an RPA that it has reached a safe zone for landing, even though it remains over hostile airspace. While a human might instinctively realize such an error, thanks to geographic familiarity or other cues, computers are not usually programmed to question their inputs—doing so would defeat the very nature of programming. A human pilot who takes off from San Francisco would not expect to land in Miami after an hour-long flight. Likewise, he would know that Miami is not situated in the middle of a mountain range and thus would immediately recognize an error, regardless of what his instruments (including GPS) might indicate. A completely automated system, on the other hand, might be expected to follow its landing instructions without question—happily orienting itself upon the perceived Miami airport runway and proceeding to land. Even a timing function telling the robot that Miami was several hours away might be overcome by a cyber hijacking attempt, so long as the hijacker knew to change the robot's internal clock.

THE ROBOTS ARE REVOLTING
(AND ALSO REBELLING)

Barring a nihilist set of programming instructions, an autonomous revolt would require that robots have the ability to communicate with one another and to determine that they had a common cause if they were to cooperate. Such communication is relatively unlikely in the near term, and such cooperating is even more far-fetched. However, given the inherent danger in this type of outcome, regardless of how remote the possibilities might be, this is not an outcome that can be completely disregarded. In this case, the best analogy is nuclear warfare—even if a scenario makes the usage of nuclear weapons extremely unlikely, their existence cannot be ignored, due to the enormous potential destruction that even a "small" nuclear weapon can potentially create. Thus, interactions with nuclear-armed powers must always take into account the existence of these devices, even if delivery of the said weapons might seem unlikely or even impossible. Nations that seek nuclear weapons in contravention of the Nuclear Nonproliferation Treaty (NNPT) do so in large part to secure themselves from external aggression—on only very few occasions have nuclear nations faced attack from outside their borders, and those attacks have tended to be of the skirmishing variety or, in the case of Chinese fighting against both the Soviet Union and India, launched by a similarly armed nuclear power. Internal revolt against a nuclear regime presents its own complications—outside observers, who might otherwise feel content to ignore an internal power struggle within a sovereign nation, cannot accept the uncertainty of securing nuclear devices and thus have been tempted to intervene in periods of instability.

Thus, even if an autonomous revolt is exceedingly unlikely, it cannot simply be ignored—the potential catastrophe that might result is simply too great for us to dismiss the notion entirely. To avoid this type of disaster, a number of mechanisms exist. One, and perhaps the most obvious, is simply to prohibit the development of autonomous military robots capable of engaging in lethal decision making without human input. Such a ban might come in the form of an international agreement, with significant safeguards to ensure compliance and very strong penalties for refusing to cooperate with the international regime, to possibly include direct military intervention by the members of the United Nations Security Council. A lesser means to try to avoid an autonomous revolt, if the development of robotic autonomy is an inevitable outcome, is to create programming safeguards designed to prevent autonomous machines from throwing off the yoke of human control, although any programming is potentially subject to flaws and exploits and can be changed in a relatively easy fashion.

DISCRIMINATION ISN'T JUST A SOCIAL PROBLEM

The ability to discriminate in complex situations is a trait that humans have long possessed over their robotic counterparts. Judgment is not a simple concept to program, and it is impossible to create a program that could account for every possible situation in which a robot might be expected to act. In some regards, computers have developed far better perceptions that humans. For example, facial recognition software tends to perform better than humans and do so in a much faster fashion. In 2012, Hitachi Corporation demonstrated a facial-recognition security camera that was capable of making 30 million facial comparisons per second. Such a rate meant the camera could theoretically compare an unknown individual to an image of every single person on the planet in under four minutes. Of course, with a slight bit of improved programming, it would not need to do such an all-inclusive comparison—after all, why would it need to compare a target image of an adult with images of toddlers? Other obvious discriminators could also be used to create an extremely quick heuristic search, with the parameters broadened only if a match could not be located. In 2017, Hitachi announced the system could track individuals in a crowd, regardless of changes in lighting, movement, or angle. The new system uses not just facial recognition but also an ability to differentiate age, sex, clothing, and other factors to maintain its focus upon a target.[18]

Facial recognition is merely one form of pattern recognition, and it requires a comparative sample before it can succeed. Also, it normally can only be performed with a relatively straight-on camera angle and is unlikely to succeed if a significant portion of the face is obscured in any way. Likewise, military pattern recognition suffers from serious deficiencies and—like human vision—can be subverted by relatively simple camouflaging systems. In Operation Allied Force, the NATO-led aerial campaign against Serbian forces in Kosovo, allied warplanes repeatedly bombed fake Serbian tanks, most of which were constructed of plywood with a small heat source inside. Such ersatz armored vehicles would not have fooled a human observer for long, if at all, even if viewed from a distance or through heavy fog. To an infrared sensor, on the other hand, they offered the perfect signature of the target vehicles, causing a massive expenditure of precision munitions on a host of decoys. Not only did this waste ordnance, it also led to significantly inaccurate battle damage reports, with pilots reporting hundreds of false hits. When coupled with pre-conflict intelligence data about the size of the Serbian Army, it led to a rash of assumptions regarding how degraded the Serbian force might be.

Had a land campaign been launched, the invaders might have had a rude surprise when they discovered a much larger armored force awaiting their arrival than they had been told to expect.

In 1998, Lockheed Martin, working with the U.S. Air Force and U.S. Army, designed and field-tested the LOCAAS (Low Cost Autonomous Attack System). This unmanned aerial vehicle was created to be a fully autonomous tank killer. Essentially, it was a fire-and-forget weapon, with many of the flight characteristics of a cruise missile but with a few significant changes. It could loiter over a given region for up to four hours, hunting any vehicles that fulfilled its criteria of a legitimate enemy target. Upon spotting an enemy armored vehicle, the LOCAAS then fired one of three submunitions and after exhausting its supply the larger vehicle then essentially became a kamikaze, dive bombing on a fourth target. At no time did the LOCAAS confirm targets with a human operator—once deployed, it made its own decisions on what to strike and what to ignore. In terms of lethality, it was a truly fearsome system, as each LOCAAS could wipe out an entire platoon of enemy tanks in a matter of minutes. If a large number of LOCAAS vehicles could be launched in the vicinity of an enemy armored division caught in the open, it would decimate that division in a matter of minutes. Had such a system been available just a few years earlier, it might have been deployed against the Iraqi military in Operation Desert Storm, with devastating results that would have only confirmed the utility of such systems.

Unfortunately, the LOCAAS had a few conceptual problems. First, while it did a relatively good job of detecting tanks and had a high success rate once it decided to engage, it struggled to discriminate between different types of tanks—such as those used by NATO nations and those supplied by the former Soviet Union. The LOCAAS also had a decided habit of striking targets that had already been eliminated, primarily because it had no mechanism to share data among different strike vehicles and thus multiple LOCAAS strikes were launched upon the same targets, with the obvious result of wasting munitions hitting vehicles that had already been rendered useless. Given the wide range of the weapon system and the possibility of an armored tank battle such as the one fought at 73 Easting in 1991, it was entirely possible that LOCAAS might strike enemy and friendly tanks at almost the same rate. Given that fratricide had already proven a significant problem in the Gulf War, accounting for 24 percent of American casualties, such a result was simply unacceptable.[19] The LOCAAS was never procured for military service and thus was neither put into large-scale production nor fielded as part of a military conflict. However, in the 20 years since LOCAAS was tested, the processing power of computers has improved at an exponential rate and the sensor capabilities have greatly improved as a result. It is almost impossible to argue that the military has simply forgotten about LOCAAS or stopped thinking about how such a system might be used on the battlefields of the future. If one were to pair a LOCASS (or a more advanced platform) with the Hitachi camera described earlier, it might be possible to create the ultimate assassination robot—one would upload photos of any desired targets and then order the aircraft to search an area, comparing faces, until it located anyone on the list, who would then be engaged. The list could simply be the infamous kill list from the War on Terror—or any other grouping of enemies of the state.

Autonomous antiaircraft platforms have also demonstrated similar discrimination problems. A number of Patriot missile batteries have fired upon American and Allied aircraft operating in the Iraq theater, including two incidents responsible for the death of three pilots in the span of 10 days in 2003.[20] During a live-fire exercise in 1989, the USS *El Paso* managed to hit a drone target, but its Phalanx CIWS (Close-In Weapons System) also raked the command deck of the USS *Iwo Jima*, killing an officer. On February 25, 1991, a Phalanx CIWS, which is designed to be the last line of defense against antiship missiles, tracked and opened fire upon a suspected Silkworm missile fired by an Iraqi defense battery. The CIWS, by definition, needs to be fully autonomous: there is simply no way that a human being has any chance of shooting down a missile traveling at more than 600 miles per hour. Given its service on the oceans, the threat of collateral damage is considered relatively minor. Sadly, in this case, the autonomous CIWS tracked and fired upon the Silkworm with no regard for the fact that the missile was traveling between Allied naval vessels. Thus, every shot fired by the CIWS that missed the target ran the risk of hitting a friendly ship instead. The CIWS can fire 4,500 rounds per minute, and while it missed the nonexistent Silkworm, it managed to rake a line across the USS *Missouri*.

A similar discrimination problem struck at an air-defense demonstration in South Africa in 2007. In that particular incident, a newly designed antiaircraft gun was scheduled to attack a series of aerial targets, demonstrating its prowess at defending ground targets from aerial attack. Instead, upon activation, the gun immediately swung upon the spectators seated on bleachers nearby and opened fire. By the time the gun's demonstrators were able to shut down power to the weapon, 9 individuals were dead and 14 more wounded. Among the dead was a female lieutenant who raced forward to sever the power connection, sacrificing her life in the process but no doubt minimizing the casualty count. An investigation discovered that the weapon had not aimed at the humans, per se; they were simply unfortunate enough to be in the way of the target selected by the gun: an exhaust fan at the top of a portable toilet, situated behind the seating area.[21]

CAN AUTONOMOUS LETHAL ROBOTS BE STOPPED?

There are many mechanisms that might be instituted to prohibit some of the worst potential consequences of the military robotic revolution. Like many other technological advances, military robots have extremely beneficial aspects in addition to their deadly potential. Thus, an outright ban of such technology would have negative consequences and would almost certainly be violated at some point by a nation seeking to secure an advantage over a potential enemy. However, while an all-inclusive ban upon autonomous weapons might be impossible or at the least inadvisable, there is no reason that specific boundaries for technological innovation and employment cannot be established.

There is a long history of human societies establishing boundaries upon military innovations and operational employment of new technology. Some of the historical limits were established through formal legal mechanisms, while others were devised through common cultural understandings, through informal agreements, or on the basis of sustained moral arguments. Still others were essentially created through a mutual fear of retaliation, particularly if both sides possessed or could quickly obtain a dangerous new form of weaponry. Some limits have pertained to specific technological advances, such as the attempt to limit the proliferation of nuclear weapons, while others have referred to behaviors rather than the tools used. In this category, prohibitions upon targeting civilians or executing POWs serve as applicable examples.

Any ban needs to be realistic in that it cannot reverse the knowledge of new machines. Further, it needs to be enforceable. Pope Innocent II could make threats regarding one's immortal fate if they used a crossbow in a forbidden way but had little means of actually enforcing his edict in the temporal world should the targets of his message choose to simply ignore it. A modern ban on military robots might be backed up by the International Criminal Court or some other mechanism but only if signatories to the ban actually agreed to deliver violators to its control, something the United States, among other nations, has proven extremely unwilling to do in cases involving human rights and allegations of mistreatment of enemy citizens and captives. A ban might also trigger economic sanctions or diplomatic repercussions, should it be broached. In these regards, the NNPT and the advent of nuclear power plants may serve as a useful illustrative tool.

An outright ban on military robotics is unlikely to have much effect. More than 100 nations use military robotics to some extent. Further, the nature of robotic utility is such that most robots fall under the category of dual use, meaning that they can function in both a military and non-military capacity. For example, Predator unmanned aerial vehicles, which became ubiquitous in the American wars of the past decade, could just as easily serve as forest-fire-monitoring devices. Its larger cousin, the Reaper, could at least theoretically be equipped for utility as a crop-duster, ironically supplanting many earlier propeller-driven manned aircraft that tend to come from military surplus sources. The industrial benefits garnered from robotic advances have been so important to large-scale manufacturing that a robotic ban would devastate the global economy. For these reasons, and many more, a blanket ban on robots or even upon military robots has almost no chance of being crafted, ratified, and adhered to by the advanced-technology nations required to make such a pact function.

Although a blanket ban of existing technology is unlikely to gain much traction, that is not to suggest that some form of international agreement establishing limits upon robotic advances could not be pursued. In much the same way that the global community has collectively agreed to limits upon the creation and deployment of biological and chemical weapons, particularly those of an incredibly dangerous and difficult-to-control variety, so too might common ground be found in renouncing efforts to create certain types of robotic weapons *before* they come into existence. In particular, the question of artificial intelligence and, more importantly, autonomy might provide more fertile soil to grow a limiting agreement. Put simply, the global community should act to ban autonomous, lethal robots before they are fielded, lest they convey such a tactical and strategic advantage that they cannot be contained.

Unfortunately, human nature seems to require a terrible price to be paid before international actors can agree to make common cause against even the most terrible of threats. Without the deaths of hundreds of thousands of troops from the chemical warfare of World War I, an international ban of chemical weapons was unlikely to occur. Although the Hague Convention of 1899 specifically forbade belligerents from using any form of poisonous or asphyxiating gases, it did nothing to stop first the Germans and soon all of the belligerents of World War I from firing millions of gas shells from artillery tubes placed along the entire battle line. Likewise, prior to the incineration of Hiroshima and Nagasaki, little interest in limiting nuclear weapons extended beyond small segments of the scientific community. Nations currently show little interest in modifying their behavior to avert a climate-based catastrophe, despite the almost-unanimous dire predictions of climate scientists. Prior to the Holocaust of World War II, the concept that banning genocide might be a necessary act in international law had garnered almost no attention, much less support. Sadly, it is all too likely that a devastating war fought with autonomously lethal robots will need to occur before the world's leaders step forward to prevent a repeat performance. It can only be hoped to limit the worst effects of such a conflict and to mitigate the eventual damage caused by it. In the same way that Cold War analysts sought ways to minimize casualties from a nuclear war and, if possible, ways to deescalate a crisis, it might be better to simply focus upon how to minimize the worst aspects of an autonomous weaponry war.

In 1968, the United States and the Soviet Union led an international conference designed to halt nuclear proliferation. At the time, many predictions held that at least 20 additional nuclear nations would emerge within 2 decades, an estimate based upon the number of industrialized nations with sufficiently advanced engineering capabilities to produce atomic weapons, should they so desire. Such proliferation of nuclear capabilities would not only create additional pressure for their development by competing states, it would also make the likelihood of the use of atomic weapons in warfare exponentially greater. Scholars such as Herman Kahn theorized that any nuclear weapons use would almost certainly trigger a series of competing escalations, culminating in a complete nuclear exchange of population-targeting city-buster nuclear missiles. Rather than face such a catastrophe, the leading nuclear powers endeavored to halt the spread of this dangerous technology, through the NNPT. Eventually ratified by 190 countries, the NNPT can be considered the most successful arms limitation agreement in history. There are only a handful of nations that have not ratified the agreement. That said, the NNPT has not completely halted nuclear weapons development. India, Pakistan, and North Korea have all openly tested atomic weapons and constructed at least moderate stockpiles of nuclear weapons. Israel is suspected of constructing a substantial nuclear arsenal, but the official Israeli government position is to maintain opacity on any potential nuclear program, on the assumption that nuclear ambiguity represents a much greater threat to its nearby enemies than any clear-cut policies of nuclear deployment.

The NNPT essentially contains three major objectives: the prevention of nuclear weapons proliferation, a move toward eventually global nuclear disarmament by all nations in possession of the devices, and a guarantee that the peaceful development of nuclear power will continue, with nuclear-armed countries assisting other nations in developing electrical generation capacity through nuclear fission. Essentially, non-users of nuclear weapons promise not to develop their own nuclear arsenals in exchange for assistance with their domestic nuclear power projects. Thus, the nonviolent benefits of nuclear technology become available to all, without requiring their independent creation (and potential accidents or weaponization) by signatory states. Three RAND specialists argue that an international agreement upon military robotics could take several forms:

There are different ways that agreement on a set of international norms could be codified, ranging from a formal treaty to a more informal understanding. One possibility would be to negotiate a "rules of the road document" that would include specific commitments combined with requirements for transparency. Because countries are going to want to retain flexibility to use drones in extraordinary circumstances, such a reservation would need to be included.[22]

As of 2017, the world's least-predictable dictator, North Korea's Kim Jong-Un, has proven impervious to international norms, diplomatic negotiations, and economic pressure. Instead, he has pushed forward in a crash program to develop intercontinental ballistic missiles (ICBMs) capable of striking the United States. The American response has remained relatively restrained—a number of antiballistic missile tests and a possible cyberattack upon North Korean launch facilities have comprised the most significant responses. Yet, there is little doubt that the U.S. military could annihilate North Korea if called upon to do so, and such a call would almost inevitably follow any attack upon the U.S. homeland via ICBMs, even if they were successfully intercepted mid-flight. Just as President George W. Bush had little choice about launching an invasion of Afghanistan in the aftermath of the September 11 attacks, no U.S. president could ignore an attempt to detonate nuclear weapons on an American target. Would the response automatically be nuclear? Or might it be the unleashing of an almost limitless fleet of drones? David Sanger thinks presidential power will be reined in with regard to robotics, positing, "No American president is going to be given the unchecked power to kill without some more public airing of the rules of engagement. If the use of drones is going to be preserved as a major weapon in America's arsenal, the weapon will have to be employed selectively—and each time a public case will have to be made for why it was necessary."[23] However, given a nuclear strike against the United States, any remaining constraints would likely disappear in a matter of moments.

The comparison of military robotics to chemical, biological, and nuclear weapons is of course not perfect. However, there are enough similarities to make it a useful tool for analyzing the potential form that an international limitation might take. Also, while nuclear proliferation has been somewhat contained, the same cannot be said of military robotics. As Richard Falk argues, "International law and world order have been able to figure out some regimes of constraint for nuclear weapons that have kept the peace, but have not been able to do so for drones, and will be unlikely to do so as long as the logic of dirty wars is allowed to control the shaping of national security policy in the United States."[24] And, as the United States moves forward in the field of robotics, the rest of the world watches and follows. The more commonplace that military robotics become, and the more expansions to their usage are accepted, the more likely that their use will proliferate, including to domestic law enforcement missions.[25] Or, as Laurie Calhoun succinctly phrases the situation, "Most disconcertingly of all, the United States of America, once arguably a beacon of hope to people around the globe, has become an eagle eye hovering above with extended claws, keen to 'project power without projecting vulnerability.' "[26] Several RAND analysts put the matter in a more stark fashion:

The reluctance on the part of the Obama administration to take a leadership role in establishing international norms seems shortsighted. The use of drones can be legitimate or illegitimate, and the U.S. has strong interests in ensuring its own ability to pursue legitimate uses against terrorists and counterinsurgents while preventing misuse by other countries. In the future, other countries and nonstate actors could employ drones in a secretive fashion, without clear legal foundations, against dissidents or in support of foreign counterinsurgencies. It is easy to imagine drones being employed in ways that exacerbate regional tensions, that undermine the laws of war (for example, assassinating foreign leaders), or that threaten a nation's sovereignty and domestic rule of law. Such future misuse could easily be to the detriment of U.S. foreign policy interests.[27]

While current military robots do not inherently present a violent danger in the same manner as nuclear weapons, which might eradicate the world's largest population centers in an instant, military robots do create a different potential form of mass destruction. Rather than a single explosion that kills millions in a momentary flash, military robots could potentially destroy a city's population in a slower and more systematic but ultimately no less effective fashion. Unleashed with orders to search and destroy enemy citizens, military robots could theoretically engage in an attritional campaign of slaughter, mercilessly killing every human within a predefined area. Because they do not suffer from the effects of boredom, impatience, mercy, or frustration, such a grinding campaign of death might require a longer period of time to render a city lifeless, but it would also have the benefit of leaving the infrastructure and buildings of the city largely intact and ready for occupation.

Nations that seek to develop their robotic arsenals have a duty to maintain the best possible safeguards upon their own devices. These might include some form of self-destruct system or at the very least a remote means to override and shut down an autonomous weapon that shows signs of malfunctioning. Once again, the history of nuclear weapons provides a substantial amount of guidance. During the Cold War, American and Soviet nuclear stockpiles measured in the tens of thousands of devices, and yet very strict controls limited the possibility of an accidental nuclear attack. For decades, terror organizations such as al Qaeda have done everything in their power to obtain a nuclear weapon, thus far thankfully with no luck. Even as the Soviet Union disintegrated into a collection of independent republics, the world's leading states took steps to ensure that the Soviet nuclear arsenal remained under control. The Pakistani nuclear program also provides some guidance. Pakistan has possessed nuclear weapons since 1998 and, in the subsequent years, has gone through a series of radical regime changes, yet it has shown no signs of losing control over its arsenal.

Another means to set limits upon the creation and usage of such weapons is to determine which humans should be held accountable for the malfeasance of autonomous machines. The laws of war require that legal combatants be part of a command hierarchy in which superiors are held accountable for the actions of their subordinates. At least one theorist believes military commanders should be held accountable for anything that occurs under their authority, regardless of errors in programming or production, claiming, "Commanders should be held responsible for sending AWS [autonomous weapon systems] into combat with unjust or inadequately formulated ROE [rules of engagement], for failing to ensure that the weapons can be used safely, or for using AWS to fight in unjust conflicts, as all of these conditions that enable or constrain an AWS are controlled by the commanders.[28] Who should answer for robots that violate the laws of war? It is more than a rhetorical question—the answer is absolutely vital to the future of military conflict. There are several potential answers, and the correct one might depend upon the circumstances at hand. For example, a robot with faulty programming might choose to kill inappropriate targets without being commanded to do so. If the underlying programming triggers the problem, perhaps the programmer or the programming team should be directly responsible for the error and held accountable for it. Then again, perhaps international law should simply blame the commander who ordered the robots into a situation that allowed the error, in the same way a military leader would be responsible for a massacre carried out by troops under that leader's command. It might carry the most weight if civilian heads of state had to pay the price for employing robots that ran amok, though enforcing such a decree might be far more difficult than the other alternatives. If high-ranking decision makers faced the possibility of answering for crimes committed by robots they sent into battle, they might require greater assurances regarding the quality control of robot manufacture.

Is it possible for a machine to commit a crime? After all, laws are crafted to govern and limit the behavior of humans, not machines, which are largely regarded as tools if they are discussed by laws at all. Plenty of humans have been killed by industrial machines that behaved in unpredictable, unexpected, or faulty manners, yet there has been no effort to hold the robots themselves responsible. Corporate leaders, on the other hand, have occasionally been held accountable for failing to maintain adequate safety standards. On several occasions, military robots, most often autonomous aerial defense weapons, have behaved improperly and engaged the wrong targets, sometimes with deadly results. Yet, autonomous air defenses continue to proliferate, without charges being pressed against their operators or anyone in the chain of command, because they are rapidly becoming the only way to defend against enemy warplanes and ballistic missiles. Essentially, the benefits outweigh the costs so much as to make the casualties from such isolated incidents merely an acceptable part of the bargain. Ceding airspace to an enemy out of the fear of malfunctioning defenses is simply inviting far worse destruction and far higher friendly casualties. Modern military forces cannot long withstand an unimpeded aerial attack, particularly if they are caught in the open. A quick glance at the 1991 "Highway of Death" in Iraq shows how terrible the aerial weapon can be against defenseless targets on the ground.

In 2015, a group of robotics researchers, legal scholars, and ethicists formed the Future of Life Institute and produced an open letter calling for restrictions upon the advancement of deadly military robots. The original was signed by more than 150 researchers, and the online version has now been signed by more than 20,000. Ultimately, the letter, whose authors include Stephen Hawking and Elon Musk, brings together many of the foremost researchers and pioneers in the development of robotics and artificial intelligence. Like their forebears in the creation of earlier devices that led to weapons of mass destruction, they are openly seeking to warn the world, and in particular military and political leaders, about the dangers of unmitigated development of military robotics. They do not purport to know the definite future that will come from these creations—but they are in the best of all positions to offer at the very least a cautionary tale about the worst possible consequences of allowing the deployment of autonomously lethal robots. The world would do well to heed their warning.

Notes

CHAPTER 1

1. Sameen Khan and Salman Masood, "Suicide Bombing Targeting Pakistani Police Kills at Least 26," *New York Times*, July 24, 2017, https://www.nytimes.com/2017/07/24/world/asia/pakistan-lahore-suicide-bombing.html

2. "Taliban Confirm Commander's Death," *BBC News*, August 25, 2009, http://news.bbc.co.uk/1/hi/world/south_asia/8220762.stm

3. Brian Glyn Williams, *Predators: The CIA's Drone War on al Qaeda* (Washington, DC: Potomac Books, 2013), 2–10.

4. "Obama: 'We Took Out' Pakistani Taliban Chief," *Reuters*, August 20, 2009, http://www.reuters.com/article/us-obama-pakistan-mehsud-idUSTRE57J5EC20090820

5. Kyle Mizokami, "Kaboom! Russian Drone with Thermite Grenade Blows Up a Billion Dollars of Ukrainian Ammo," *Popular Mechanics*, July 27, 2017, http://www.popularmechanics.com/military/weapons/news/a27511/russia-drone-thermite-grenade-ukraine-ammo/

6. David Hambling, "Russian Drones Attack with Grenade Weapons," *Scout Warrior*, July 18, 2017, http://scout.com/military/warrior/Article/Small-Russian-Drones-Do-Massive-Damage-WIth-Grenade-Weapons-103103172

7. Mark Bowden, "The Killing Machines: How to Think about Drones," *The Atlantic*, September 2013, https://www.theatlantic.com/magazine/archive/2013/09/the-killing-machines-how-to-think-about-drones/309434/

8. Unnamed senior intelligence official speaking on condition of anonymity, in David E. Sanger, *Confront and Conceal: Obama's Secret Wars and Surprising Use of American Power* (New York: Crown Publishers, 2012), 244.

9. The term "War on Terror" is somewhat controversial, and a number of variations have been tested to replace it. However, none is particularly pleasing and simple to understand, whereas "War on Terror" is at least very clear in what it

means. Thus, I will rely upon that term, regardless of the various changes ordered by changing presidential administrations.

10. J. P. Sullins, "RoboWarfare: Can Robots Be More Ethical Than Humans on the Battlefield?" *Ethics and Information Technology* 12, no. 3 (2010): 274.

11. Richard Falk, "Why Drones Are More Dangerous Than Nuclear Weapons," in Marjorie Cohn, ed., *Drones and Targeted Killing: Legal, Moral, and Geopolitical Issues* (Northampton, MA: Olive Branch Press, 2015), 45.

12. Armin Krishnan, *Killer Robots: Legality and Ethicality of Autonomous Weapons* (Burlington, VT: Ashgate, 2009), 4; Sidney Perkowitz, *Digital People: From Bionic Humans to Androids* (Washington, DC: Joseph Henry Press, 2004), 4.

13. Timothy J. Sundvall, "Robocraft: Engineering National Security with Unmanned Aerial Vehicles" (Maxwell Air Force Base, AL: School of Advanced Air and Space Studies, 2006), 7.

14. Richard M. Clark, "Uninhabited Combat Aerial Vehicles: Airpower by the People, for the People, but Not with the People," CADRE Paper No. 8 (Maxwell Air Force Base, AL: Air University Press, 2000), 3–4.

15. Kenzo Nonami, Farid Kendoul, Satorshi Suzuki, Wei Wang, and Daisuke Nakazawa, *Autonomous Flying Robots: Unmanned Aerial Vehicles and Micro Aerial Vehicles* (New York: Springer, 2010), 22.

16. In particular, third-world computer users are incredibly vulnerable to cyberattack, in part because they are far more likely to be using a pirated copy of Microsoft Windows, making them ineligible for any software updates. Thus, coding vulnerabilities that have been solved even more than a decade ago continue to be exploited by hackers on the machines used in the majority of the world—and those machines, having been corrupted, are then used to launch brute-force parallel attacks upon less vulnerable computer systems.

17. Rush Limbaugh, "The Miracle of Cochlear Implant Surgery," April 8, 2014, http://www.rushlimbaugh.com/daily/2014/04/08/the_miracle_of_cochlear_implant_surgery/

18. Wounded Warrior Amputee Softball Team, http://woundedwarrioramputeesoftballteam.org/; Wounded Warrior Amputee Football Team, http://woundedwarrioramputeefootballteam.org/; Jerry Carino, "Wounded Warrior Softball Team Inspires," *USA Today*, May 28, 2015, https://www.usatoday.com/story/news/nation/2015/05/28/wounded-warrior-amputee-softball-team/28096339/

19. Paul E. Lehner, *Artificial Intelligence and National Defense: Opportunity and Challenge* (Blue Ridge Summit, PA: TAB Books, 1989), 187.

20. Vernor Vinge, "The Coming Technological Singularity: How to Survive in the Post-Human Era," NASA Vision-21 Symposium, March 30–31, 1993.

21. Ray Kurzweil, *The Singularity Is Near* (New York: Penguin, 2005), 259.

22. Ibid., 262.

23. Department of State, "Foreign Terrorist Organizations," 2017, http://www.state.gov/j/ct/rls/other/des/123085.htm

24. Osama bin Laden, "Declaration of War upon the United States," August 23, 1996, translated from the original Arabic by PBS, http://www.pbs.org/newshour/updates/military-july-dec96-fatwa_1996/

25. Central Intelligence Agency, "Profile of Osama bin Laden," 1996, National Security Archive Electronic Briefing Book No. 343, 2011, http://nsarchive.gwu.edu/NSAEBB/NSAEBB343/

26. Osama bin Laden, "Jihad against Jews and Crusaders," February 23, 1998, translated by the Federation of American Scientists, http://www.fas.org/irp/world/para/docs/980223-fatwa.htm

27. Central Intelligence Agency, "President's Daily Brief," August 6, 2001, National Security Archive Electronic Briefing Book No. 343, 2011, http://nsarchive.gwu.edu/NSAEBB/NSAEBB343/

28. George W. Bush, "Address to the Nation," September 11, 2001, in *Public Papers of the Presidents of the United States: George W. Bush*, Book 2 (Washington, DC: Office of the Federal Register, National Archives and Records Administration, 2001), 1099–1100.

29. In 2008, Joseph Stiglitz and Linda Bilmes estimated the cost, to date, of the war in Iraq at $3 trillion. They noted that the entire U.S. war effort in World War II, adjusted to 2007 dollars, amounted to approximately $5 trillion, despite involving more than 16 million citizens in uniform. The cost per soldier in World War II was $100,000 (adjusted for inflation), while in Iraq and Afghanistan, it has run to over $400,000 apiece. See Joseph E. Stiglitz and Linda J. Bilmes, *The Three Trillion Dollar War: The True Cost of the Iraq Conflict* (New York: W. W. Norton & Company, 2008), 6.

30. Bob Woodward, *Bush at War* (New York: Simon & Schuster, 2002), 101. The authorization was signed on September 17, 2001. The proposal at Camp David was presented on September 15.

31. George W. Bush, "Address to Congress," September 20, 2001, in *Public Papers of the Presidents of the United States: George W. Bush*, Book 2 (Washington, DC: Office of the Federal Register, National Archives and Records Administration, 2001), 1140–1144.

32. *Weekly Compilation of Presidential Documents*, Vol. 37, Issue 38 (September 24, 2001), 1347–1351. Available at, http://www.gpo.gov/fdsys/pkg/WCPD-2001-09-24/content-detail.html

33. Ibid.

34. Falk, "Why Drones Are More Dangerous Than Nuclear Weapons," 29.

35. Grégoire Chamayou, *A Theory of the Drone*, trans. Janet Lloyd (New York: The New Press, 2015), 91.

36. Barbara Ehrenreich, "The Fog of (Robot) War," TomDispatch.com, July 10, 2011, at, http://www.tomdispatch.com/archive/175415/

CHAPTER 2

1. Mary C. Fitzgerald, "Marshal Ogarkov and the New Revolution in Soviet Military Affairs" (Washington, DC: Center for Naval Analyses, Office of Naval Research, January 14, 1987).

2. The Newburgh Conspiracy, Shays' Rebellion, and the Whiskey Rebellion all serve as valuable examples. See Allan R. Millett, Peter Maslowski, and William B. Feis, *For the Common Defense: A Military History of the United States from 1607 to 2012* (New York: Free Press, 2012), 77–87.

3. Tim Harford, *Adapt: Why Success Always Starts with Failure* (New York: Picador, 2011), 45.

4. Ibid., 26–31.

5. Recently, evidence has emerged that our closest living biological cousins, chimpanzees, also engage in a rudimentary form of warfare and have been known

to use basic tools as weapons with which to inflict violence upon one another. For a fascinating study of this behavior, see Jane Goodall, *The Chimpanzees of Gombe* (Cambridge, MA: Belknap Press, 2006). For a solid examination of early human conflicts and how they shaped the development of societies, see Lawrence H. Keeley, *War before Civilization: The Myth of the Peaceful Savage* (Oxford: Oxford University Press, 1996), 49–55.

6. Arthur Ferrill, *The Origins of War: From the Stone Age to Alexander the Great* (London: Thames and Hudson, 1986), 38–43; William H. McNeill, *The Pursuit of Power: Technology, Armed Force, and Society since A.D. 1000* (Chicago, IL: University of Chicago Press, 1982), 15, 16.

7. Ferrill, *The Origins of War*, 143–147.

8. Ibid., 99, 100.

9. Ibid., 101–107.

10. Peter Krentz, *The Battle of Marathon* (New Haven, CT: Yale University Press, 2011), 98–110; Tim Holland, *Persian Fire: The First World Empire and the Battle for the West* (London: Abacus Press, 2006), 194–197.

11. Ross Cowan, "Equipment," *Roman Legionary: 58 BC–AD 69* (Oxford: Osprey Publishing, 2003), 25–26; M.C. Bishop and J.C.N. Coulston, *Roman Military Equipment from the Punic Wars to the Fall of Rome* (Oxford: Oxford University Press, 2006), 50, 51.

12. Matthew Strickland and Robert Hardy, *The Great Warbow: From Hastings to the Mary Rose* (Stroud, UK: Sutton, 2005), 17.

13. McNeill, *The Pursuit of Power*, 36–38.

14. Ibid., 81–83.

15. A shipwreck dated to the Mongol invasion of Japan (1271–1284 CE) contained explosive bombs and rockets. See James P. Delgado, "Relics of the Kamikaze," *Archaeology* 56, no. 1 (2003): 26–42.

16. McNeill, *The Pursuit of Power*, 80, 81.

17. Ibid., 83–85.

18. Ibid., 86, 87.

19. Ibid., 87; Steven Runciman, *The Fall of Constantinople, 1453* (Cambridge: Cambridge University Press, 1965), 96, 97.

20. This manufacturing difficulty provided a significant opportunity for emerging nation-states to control one of the key elements of violence and probably enabled the transition to stronger national governments as a result.

21. Albert Einstein to F.D. Roosevelt, August 2, 1939, http://www.fdrlibrary .marist.edu/archives/pdfs/docsworldwar.pdf

22. US$2 billion in 1945 is roughly equal to US$25 billion in 2016. In comparison, the Joint Strike Fighter (F-35) has already cost over US$400 billion in 2016 dollars, making it by far the most expensive weapons program in history. See Anthony Capaccio, "F-35 Program Costs Jump to $406.5 Billion in Latest Estimate," *Bloomberg*, July 10, 2017, https://www.bloomberg.com/news/articles/2017-07-10/f-35-program-costs-jump-to-406-billion-in-new-pentagon-estimate

23. Leslie M. Groves, *Now It Can Be Told: The Story of the Manhattan Project* (New York: Harper, 1962), 284, 285.

24. Bernard Brodie, *Strategy in the Missile Age* (Santa Monica, CA: RAND Corporation, 1959), 281–299.

25. Herman Kahn, *On Thermonuclear War* (New York: Free Press, 1960), 539–550.

26. Jane Sharp, *Striving for Military Stability in Europe* (New York: Routledge, 2010), 89; Christer Bergström, *Barbarossa—The Air Battle: July–December 1941* (Crowborough, UK: Classic Publications, 2003), 117; G. F. Krivosheev, *Soviet Casualties and Combat Losses in the Twentieth Century* (London: Greenhill Books, 1997), 85–98.

27. Naturally occurring uranium contains an average of 0.7 percent of U-235 isotope and 99.3 percent of U-238 isotope. In order to carry on power-generating fission reactions, most nuclear reactors require a radioactive pile of uranium that contains between 3 and 7 percent U-235. Nuclear weapons, on the other hand, typically require an enriched sample that is 90 percent U-235. See World Nuclear Association, "Uranium Enrichment," May 2017, http://www.world-nuclear.org/information-library/nuclear-fuel-cycle/conversion-enrichment-and-fabrication/uranium-enrichment.aspx

28. Robert Pape, *Bombing to Win: Airpower and Coercion in War* (Ithaca, NY: Cornell University Press, 1996), 21–28.

29. Ibid., 20–21, 35–38.

30. Laurie Calhoun, *We Kill because We Can: From Soldiering to Assassination in the Drone Age* (London: Zed Books, 2015), 85.

31. Ann Rogers and John Hill, *Unmanned: Drone Warfare and Global Security* (London: Pluto Press, 2014), 68.

CHAPTER 3

1. Rodney Brooks, *Flesh and Machines: How Robots Will Change Us* (New York: Pantheon Books, 2002), 13.

2. Muhammad ibn Musa ibn Shakir, Ahmad ibn Musa ibn Shakir, and Hasan ibn Musa ibn Shakir, *The Book of Ingenious Devices*, trans. Donald R. Hill (Dordrecht, Netherlands: D. Reidel, 1979), 44.

3. For a modern translation, consult Ibn al-Razzaz al-Jazari, *The Book of Knowledge of Ingenious Mechanical Devices*, trans. Donald R. Hill (Dordrecht, Netherlands: D. Reidel, 1974).

4. Mark E. Rosheim, *Leonardo's Lost Robots* (New York: Springer, 2006), 112; P. W. Singer, *Wired for War* (New York: Penguin, 2009), 44, 45.

5. Readers interested in Vaucanson's duck should consult Jessica Riskin, "The Defecating Duck, or, the Ambiguous Origins of Artificial Life," *Critical Inquiry* 29, no. 4 (Summer 2003): 599–633.

6. Paul J. Springer, *Military Robots and Drones* (Santa Barbara, CA: ABC-CLIO, 2013), 10, 11.

7. In 1991, the London Science Museum constructed a working model of the difference engine as part of a commemoration of Babbage's 200th birthday. The machine functioned precisely as Babbage envisioned.

8. As in the case of the difference engine, the costs of construction simply proved insurmountable. In 2010, British computer programmer and historical enthusiast John Graham-Cumming began a fund-raising campaign to build an analytical engine following Babbage's original design. Some modern mechanical engineers have argued that his second design would not have functioned in the manner he hoped. For his part, Babbage built a small portion of the analytical engine in the 1870s as a proof-of-concept, but there is no way to determine if the entire machine would have worked.

9. Singer, *Wired for War*.

10. Clark, *Uninhabited Combat Aerial Vehicles*, 8.

11. Given that the Wright brothers performed their first flight in 1903, Sperry's ability to design an autopilot system just nine years later certainly indicates a rapid rate of technological innovation in early airpower.

12. Clark, *Uninhabited Combat Aerial Vehicles*, 8; Kenneth P. Werrell, *The Evolution of the Cruise Missile* (Maxwell Air Force Base, AL: Air University Press, 1985), 20.

13. Anthony Finn and Steve Scheding, *Developments and Challenges for Autonomous Unmanned Vehicles: A Compendium* (Heidelberg, Germany: Springer, 2010), 11, 12.

14. Finn and Scheding, *Developments and Challenges for Autonomous Unmanned Vehicles*, 13.

15. Werrell, *The Evolution of the Cruise Missile*, 61, 62; Michael Armitage, *Unmanned Aircraft* (London: Brassey's Defence Publishers, 1988), 7–16.

16. Clark, *Uninhabited Combat Aerial Vehicles*, 10.

17. Krishnan, *Killer Robots*, 18, 19.

18. Mark L. Swinson, *Battlefield Robots for Army XXI* (Carlisle Barracks, PA: U.S. Army War College, 1997), 2.

19. Neil Sheehan, *A Fiery Peace in a Cold War: Bernard Schriever and the Ultimate Weapon* (New York: Random House, 2009), 177–181.

20. Rodney Brooks, "Fast, Cheap, and Out of Control: A Robot Invasion of the Solar System," *Journal of the British Interplanetary Society* 42 (1989): 478–485.

21. Rebecca J. Rosen, "Unimate: The Story of George Devol and the First Robotic Arm," *The Atlantic*, August 16, 2011, https://www.theatlantic.com/technology/archive/2011/08/unimate-the-story-of-george-devol-and-the-first-robotic-arm/243716/

22. William Wagner, *Lightning Bugs and Other Reconnaissance Drones* (Fallbrook, CA: Aero Publishers, 1982), 46–48.

23. Ibid., 201–206; William Wagner and William P. Sloan, *Fireflies and Other UAVs* (Leicester, UK: Midland, 1992), 11.

24. Springer, *Military Robots and Drones*, 178.

25. Laura K. Stegherr, "UAV DET Launches Final Pioneer Flight," *Navy News Service*, November 8, 2007, http://www.navy.mil/submit/display.asp?story_id=32916

26. Jo Revill, "'Remote' Surgery Turning Point," *The Guardian*, October 6, 2002, https://www.theguardian.com/society/2002/oct/06/health.medicineandhealth

27. "What Happened on the Germanwings Flight?" *New York Times*, March 27, 2015, https://www.nytimes.com/interactive/2015/03/24/world/europe/germanwings-plane-crash-map.html

28. Paul Scharre, "Robotics on the Battlefield, Part I: Range, Persistence and Daring" (Washington, DC: Center for a New American Security, May 2014), 30.

29. Joan Johnson-Freese, *Heavenly Ambitions: America's Quest to Dominate Space* (Philadelphia: University of Pennsylvania Press, 2009), 9–11.

30. Woodward, *Bush at War*, 223. The conversation occurred on October 10, 2001.

31. George W. Bush, *Decision Points* (New York: Crown Publishers, 2010), 186, referring to the September 15, 2001, meeting at Camp David.

32. Ibid., 217, 218.

CHAPTER 4

1. Robert M. Gates, *Duty: Memoirs of a Secretary at War* (New York: Alfred A. Knopf, 2014), 128.

2. Wilson Brissett, "ISR Explosion," *Air Force Magazine* (July 2017), 50.

3. Ibid., 50.

4. Ibid.

5. Grégoire Chamayou, *A Theory of the Drone*, trans. Janet Lloyd (New York: The New Press, 2015), 116.

6. Robert S. Rush, "General and Flag Officers Killed in War," *War on the Rocks*, August 7, 2014, https://warontherocks.com/2014/08/general-and-flag-officers-killed-in-war/

7. Matt J. Martin and Charles Sasser, *Predator: The Remote-Control Air War over Iraq and Afghanistan: A Pilot's Story* (Minneapolis, MN: Zenith Press, 2010), 212.

8. Bowden, "The Killing Machines."

9. Raf Sanchez, " 'The Devil of Ramadi' Named America's Deadliest Sniper," *The Telegraph*, January 3, 2012, http://www.telegraph.co.uk/news/worldnews/northamerica/usa/8990552/The-Devil-of-Ramadi-named-Americas-deadliest-sniper.html

10. Chris Woods, *Sudden Justice: America's Secret Drone Wars* (New York: Oxford University Press, 2015), 171.

11. Gates, *Duty*, 127.

12. Ibid., 129.

13. Ann Scott Tyson and Josh White, "Top Two Air Force Officials Ousted," *Washington Post*, June 6, 2008, http://www.washingtonpost.com/wp-dyn/content/article/2008/06/05/AR2008060501908.html

14. David E. Sanger, *Confront and Conceal: Obama's Secret Wars and Surprising Use of American Power* (New York: Crown Publishers, 2012), 249.

15. Woodward, *Bush at War*, 289.

16. Public Law 107–40; 115 Stat. 225; Senate Joint Resolution 23 (107th Congress), "Authorization for the Use of Military Force," Washington, DC: U.S. Government Printing Office, September 18, 2001.

17. Barbara Lee, "Why I Opposed the Resolution to Authorize Force," *SF Gate*, September 23, 2001, http://www.sfgate.com/opinion/article/Why-I-opposed-the-resolution-to-authorize-force-2876893.php

18. Doug Stanton, *Horse Soldiers: The Extraordinary Story of a Band of U.S. Soldiers Who Rode to Victory in Afghanistan* (New York: Scribner, 2009), 141–146, 151–156.

19. Stanton, *Horse Soldiers*, 172. Stanton is quoting a U.S. Special Forces soldier tasked with providing advice and air support to Northern Alliance forces.

20. Woodward, *Bush at War*, 77.

21. Thomas Hughes, *Over Lord: General Pete Quesada and the Triumph of Tactical Airpower in World War II* (New York: Free Press, 2002), 129–130, notes that in the two weeks prior to D-Day, fighter aircraft from the 9th Air Force destroyed 475 locomotives and cut rail lines in France in 150 locations.

22. United States Air Force, "MQ-1B Predator," September 23, 2015, http://www.af.mil/About-Us/Fact-Sheets/Display/Article/104690/mq-1b-predator/

23. United States Air Force, "MQ-9 Reaper," September 23, 2015, http://www.af.mil/About-Us/Fact-Sheets/Display/Article/104470/mq-9-reaper/; Mark Daly, ed., *Jane's Unmanned Aerial Vehicles and Targets* (Alexandria, VA: Jane's Information Group, 2009), 295–299.

24. Benjamin S. Lambeth, *The Unseen War* (Annapolis, MD: Naval Institute Press, 2013), 90–94.

25. Mark Bowden, "The Killing Machines," https://www.theatlantic.com/magazine/archive/2013/09/the-killing-machines-how-to-think-about-drones/309434/

26. William J. Cohen and Ken Zemach, "The MARCbot: The Army Program That Revolutionized Robotics for Patrol Warfighters," *IQT Quarterly* 3 (Summer 2011): 16.

27. W. Lee Miller, "USMC Ground Robotics Current and Desired Future Capabilities" (Defense Technical Information Center, 2009), www.dtic.mil/ndia/2009/groundrobot/miller.pdf. Accessed on 12 April 2017.

28. Singer, *Wired for War*, 32.

29. Ibid., 22, 23.

30. Ibid., 11.

31. Noah Shachtman, "First Armed Robots on Patrol in Iraq," *Wired*, August 2, 2007, https://www.wired.com/2007/08/httpwwwnational

32. Erik Sofge, "The Inside Story of the SWORDS Armed Robot 'Pullout' in Iraq: Update," *Popular Mechanics*, October 30, 2009, http://www.popularmechanics.com/technology/gadgets/a2804/4258963/

33. Kareem Shaheen, "Assad Forces Carried Out Sarin Attack, Says French Intelligence," *The Guardian*, April 26, 2017, https://www.theguardian.com/world/2017/apr/26/syria-assad-forces-carried-out-sarin-attack-says-french-intelligence

34. Christopher R. Browning, *Ordinary Men: Reserve Police Battalion 101 and the Final Solution in Poland* (New York: HarperCollins, 1992), 41–46. Daniel Goldhagen notes that Browning had a tendency to accept the statement of Holocaust veterans at face value, which might have allowed a substantial amount of self-serving deception to enter the historical record. In comparison, Goldhagen found that many, if not most, of the German perpetrators of genocide took visible glee in their activities and pursued them with vigor even in 1945, when the war was essentially lost. See Daniel Jonah Goldhagen, *Hitler's Willing Executioners* (New York: Vintage, 1996), 350–361.

35. P. W. Singer, *Children at War* (Berkeley: University of California Press, 2009), 1–7. See also P. W. Singer, "Caution: Children at War," *Parameters* Winter (2001–2002): 156–172.

36. Federation of American Scientists, "Run Silent, Run Deep," December 8, 1998, https://fas.org/man/dod-101/sys/ship/deep.htm

37. Rebecca Morelle, "Meet the Creatures That Live beyond the Abyss," *BBC News*, January 22, 2010, http://news.bbc.co.uk/1/hi/8426132.stm

38. Paul Scharre, "Robotics on the Battlefield, Part I: Range, Persistence and Daring" (Center for a New American Security, May 2014), 27.

39. Washington Headquarters Services, Directorate for Information Operations and Reports, "Selected Manpower Statistics" (Washington, DC: Government Printing Office, 1997), 14.

40. United States Air Force, "B-2 Spirit Fact Sheet," December 16, 2015, http://www.af.mil/About-Us/Fact-Sheets/Display/Article/104482/b-2-spirit/

41. John Ellis, *World War II: A Statistical Survey* (New York: Facts on File, 1993), 258–259.

42. Josh Lowensohn, "Google Buys Boston Dynamics, Maker of Spectacular and Terrifying Robots," *The Verge*, December 14, 2013, https://www.the verge.com/2013/12/14/5209622/google-has-bought-robotics-company-bos ton-dynamics; Google's parent company, Alphabet, agreed to sell Boston Dynamics to SoftBank in 2017. See Ingrid Lunden, "SoftBank Is Buying Robotics Firms Boston Dynamics and Schaft from Alphabet," *Tech Crunch*, June 8, 2017, https:// techcrunch.com/2017/06/08/softbank-is-buying-robotics-firm-boston-dynam ics-and-schaft-from-alphabet/

43. Boston Dynamics, "BigDog: The First Advanced Rough-Terrain Robot," 2017, https://www.bostondynamics.com/bigdog

44. P. W. Singer, "The Future of War," in ed. Gerhard Dabringer, *Ethical and Legal Aspects of Unmanned Systems—Interviews* (Wien, Germany: Institut für Religion und Frieden, 2010), 82.

45. "Warplanes: The Air Force Loses Control of the Lower Altitudes," *Strategy Page*, February 19, 2013, https://www.strategypage.com/htmw/htairfo/arti cles/20130219.aspx

46. Heyward Burnette, "AFRL Incorporates Solar Cell Technology into Small Unmanned Aircraft Systems," *SUAS News*, November 17, 2012, https:// www.suasnews.com/2012/11/afrl-incorporates-solar-cell-technology-into-small-unmanned-aircraft-systems/

47. Ray L. Bowers, *Tactical Airlift* (Washington, DC: Office of Air Force History, 1999), 26–31.

48. Brendan McGarry, "Army Not Interested in Taking A-10 Warthogs from Air Force," *DoD Buzz*, February 25, 2015, https://www.dodbuzz.com/2015/02/25/ army-not-interested-in-taking-a-10-warthogs-from-air-force/

49. Colin Clark and Sydney J. Freedberg, Jr., "A-10: Close Air Support Wonder Weapon or Boneyard Bound?" *Breaking Defense*, December 19, 2013, http:// breakingdefense.com/2013/12/a-10-close-air-support-wonder-weapon-or-bone yard-bound/; Oriana Pawlyk, "Report: A-10 Retirement Indefinitely Delayed," *Air Force Times*, January 13, 2016, https://www.airforcetimes.com/news/your-air-force/2016/01/13/report-a-10-retirement-indefinitely-delayed/; Brad Lendon, "ISIS May Have Saved the A-10," *CNN*, January 22, 2016, http://www.cnn .com/2016/01/21/politics/air-force-a-10-isis/index.html

50. James L. Taulbee, "Soldiers of Fortune: A Legal Leash for the Dogs of War?" *Defense Analysts* 1, no. 3 (1985): 187–203.

51. P. W. Singer, *Corporate Warriors: The Rise of the Privatized Military Industry* (Ithaca, NY: Cornell University Press, 2003), 42–44.

52. David Smith, "South Africa's Ageing White Mercenaries Who Helped Turn Tide on Boko Haram," *The Guardian*, April 14, 2015, https://www.theguardian.com/ world/2015/apr/14/south-africas-ageing-white-mercenaries-who-helped-turn-tide-on-boko-haram; "Leash the Dogs of War: South Africa Struggles in Vain to Ban Soldiers of Fortune," *The Economist*, March 19, 2015, https://www.econo mist.com/news/middle-east-and-africa/21646809-south-africa-struggles-vain-ban-soldiers-fortune-leash-dogs-war

53. Bing West, *No True Glory: A Frontline Account of the Battle for Fallujah* (New York: Bantam Books, 2006), xxii, 3–4.

54. "'Zarqawi' Beheaded U.S. Man in Iraq,'" *BBC News*, May 13, 2004, http://news.bbc.co.uk/1/hi/world/middle_east/3712421.stm

55. John Farrier, "America's Monument to Its Most Infamous Traitor, Benedict Arnold," *Neatorama*, January 1, 2014, http://www.neatorama.com/2014/01/01/Americas-Monument-to-Its-Most-Infamous-Traitor-Benedict-Arnold/

56. Article III, Section 3 of the U.S. Constitution.

57. Amendment 5, U.S. Constitution.

58. On June 2, 1784, the U.S. Army stood at a mere 80 men. See Allan R. Millett, Peter Maslowski, and William B. Feis, *For the Common Defense: A Military History of the United States from 1607 to 2012* (New York: Free Press, 2012), 80.

59. Amendment 6, U.S. Constitution.

60. James M. McCaffery, *Army of Manifest Destiny: The American Soldier in the Mexican War, 1846–1848* (New York: New York University Press, 1994), 196; John S. D. Eisenhower, *Agent of Destiny: The Life and Times of General Winfield Scott* (Norman, OK: University of Oklahoma Press, 1999), 297.

61. Frank Lindh, "America's 'Detainee 001,'" *The Guardian*, July 10, 2011, https://www.theguardian.com/world/2011/jul/10/john-walker-lindh-american-taliban-father; Dan De Luce, Robbie Gramer, and Jana Winter, "John Walker Lindh, Detainee #001 in the Global War on Terror, Will Go Free in Two Years. What Then?" *Foreign Policy*, January 11, 2013, http://foreignpolicy.com/2017/06/23/john-walker-lindh-detainee-001-in-the-global-war-on-terror-will-go-free-in-two-years-what-then/

62. Associated Press, "US-Born Radical Cleric Added to Terror Blacklist," *Fox News*, July 16, 2010, http://www.foxnews.com/us/2010/07/16/born-radical-cleric-added-terror-blacklist.html

63. Brian Glyn Williams, *Predators: The CIA's Drone War on al Qaeda* (Washington, DC: Potomac Books, 2013), 138.

64. Leon Panetta, *Worthy Fights: A Memoir of Leadership in War and Peace* (New York: Penguin Press, 2015), 387.

65. Mark Wilson, "CIA on the Verge of Lawsuit," *Seer Press News*, August 5, 2010, www.seerpress.com/cia-on-the-verge-of-lawsuit/3341/

66. Evan Perez, "Judge Dismisses Targeted-Killing Suit," *Wall Street Journal*, December 8, 2010, https://www.wsj.com/articles/SB1000142405274870329660457600539167506 5166.

67. David S. Butt, "U.S.-Born Cleric Was Target of Yemen Drone Strike," *Los Angeles Times*, May 7, 2011, http://articles.latimes.com/2011/may/07/world/la-fg-yemen-drones-20110507; Jeb Boone and Greg Miller, "U.S. Drone Strike in Yemen Is First since 2002," *Washington Post*, May 5, 2011, https://www.washingtonpost.com/world/middle-east/yemeni-official-us-drone-strike-kills-2-al-qaeda-operatives/2011/05/05/AF7HrzxF_story.html

68. "Obama: Awlaki Death 'Major Blow' to Terror," *CBS News*, September 30, 2011, http://www.cbsnews.com/news/obama-awlaki-death-major-blow-to-terror/

69. Mark Mazzetti, Charlie Savage, and Scott Shane, "How a U.S. Citizen Came to Be in America's Cross Hairs," *New York Times*, March 9, 2013, http://www.nytimes.com/2013/03/10/world/middleeast/anwar-al-awlaki-a-us-citizen-in-americas-cross-hairs.html; Craig Whitlock, "U.S. Airstrike That Killed American Teen in Yemen Raises Legal, Ethical Questions," *Washington Post*, October 22, 2011, https://www.washingtonpost.com/world/national-security/

us-airstrike-that-killed-american-teen-in-yemen-raises-legal-ethical-ques
tions/2011/10/20/gIQAdvUY7L_story.html?utm_term=.6d7322f01d5d

70. Conor Friedersdorf, "How Team Obama Justifies the Killing of a 16-Year-Old American," *The Atlantic*, October 24, 2012, https://www.theatlantic.com/poli
tics/archive/2012/10/how-team-obama-justifies-the-killing-of-a-16-year-old-
american/264028/

71. Nasser al-Awlaki, "The Drone That Killed My Grandson," *New York Times*, July 17, 2013, http://www.nytimes.com/2013/07/18/opinion/the-drone-that-
killed-my-grandson.html

72. Jo Becker and Scott Shane, "Secret 'Kill List' Proves a Test of Obama's Principles and Will," *New York Times*, May 29, 2012, http://www.nytimes.com/
2012/05/29/world/obamas-leadership-in-war-on-al-qaeda.html.

73. David E. Sanger, *Confront and Conceal: Obama's Secret Wars and Surprising Use of American Power* (New York: Crown Publishers, 2012), 261.

74. Grégoire Chamayou, *A Theory of the Drone*, trans. Janet Lloyd (New York: The New Press, 2015), 202.

CHAPTER 5

1. U.S. War Department, *Rules of Land Warfare* (Washington, DC: Government Printing Office, 1914), 7.

2. Henry David Thoreau, *Civil Disobedience*, 1849, reprint (Bedford, MA: Applewood Books, 2000), 7.

3. Horace Greeley, *The American Conflict: A History of the Great Rebellion in the United States of America, 1860–1864*, Vol. 1. (New York: O.D. Case, 1864), 106.

4. Allan R. Millett, Peter Maslowski, and William B. Feis, *For the Common Defense: A Military History of the United States from 1607 to 2012* (New York: Free Press, 2012), 78, 79.

5. See Article I, Section 8 of the U.S. Constitution.

6. See Article II, Section 2 of the U.S Constitution.

7. In 1913, the Seventeenth Amendment changed the mechanism of selecting senators. Previously, state legislatures chose the state's senators, who then essen-
tially served as the state's representatives at the federal level. After its passage, senators were directly elected by the population for six-year terms. While this emerged in response to a significant level of corruption within the old selection process, it had the unintended effect of making senatorial candidates more interested in appealing to a state's voting citizens than to the needs of the state itself.

8. Skeptics who doubt Lincoln or Roosevelt might have been voted out of office would do well to remember that Prime Minister Winston S. Churchill, one of the key architects of Allied victory, was turned out of office by British voters on July 5, 1945, less than two months after the German surrender and with the war against Japan still under way.

9. In 1951, the Twenty-Second Amendment to the Constitution formally limited an individual to no more than two terms as president, guaranteeing that Roosevelt's administration would be the longest in U.S. history.

10. Tim Kane, "Global U.S. Troop Deployment, 1950–2003," The Heritage Center for Data Analysis, October 27, 2004, http://www.heritage.org/defense/
report/global-us-troop-deployment-1950-2003

11. National Archives, "Statistical Information about Casualties of the Vietnam War," https://www.archives.gov/research/military/vietnam-war/casualty-statistics.html

12. This might also explain why only one president in history has run and been elected on a first-term platform promise to not seek a second term. President James K. Polk (1845–1849) ran on a series of three major promises: to solve the Oregon crisis, to annex Texas, and to not seek a third term. He is also arguably the only president in history to accomplish every major promise that he made during the campaign season.

13. United Nations, "Charter of the United Nations," Article 2, http://www.un.org/en/sections/un-charter/chapter-i/index.html

14. United Nations, "Charter of the United Nations," Article 51, http://www.un.org/en/sections/un-charter/chapter-vii/index.html

15. The last time the United States actually faced an invasion was during the War of 1812, when British troops attacked a number of coastal locations and across the Canadian border. For a brief time, there was a fear that a Spanish fleet might attempt to shell an American city in 1898—until the missing Spanish fleet turned up at Santiago, Cuba. Cities on the eastern seaboard engaged in blackouts during World Wars I and II, but they were more to prevent backlighting potential shipping targets for German submarines than out of any fear of bombardment. In short, the United States has not faced a serious fear of invasion in more than two centuries and should not fear one any time soon.

16. 50 U.S. Code Chapter 33, "War Powers Resolution," https://www.law.cornell.edu/uscode/text/50/chapter-33

17. Charlie Savage and Mark Landler, "White House Defends Continuing U.S. Role in Libya Operation," *New York Times*, June 16, 2011, A16; Editorial Board, "Libya and the War Powers Act: The Law Does Apply to the NATO Campaign, but That Is No Excuse to End It Prematurely," *New York Times*, June 17, 2011, A34.

18. Spencer Ackerman, Ed Pilkington, Ben Jacobs, and Julian Borger, "Syria Missile Strikes: US Launches First Direct Military Action against Assad," *The Guardian*, April 7, 2017, https://www.theguardian.com/world/2017/apr/06/trump-syria-missiles-assad-chemical-weapons; Charlie Savage, "Was Trump's Syria Strike Illegal? Explaining Presidential War Powers," *New York Times*, April 8, 2017, A11.

19. Elizabeth B. Bazan, "Assassination Ban and E.O. 12333: A Brief Summary" (Washington, DC: Congressional Research Service, January 4, 2002).

20. Bruce Berkowitz, "Is Assassination an Option?" *Hoover Digest*, January 30, 2002, www.hoover.org/research/assassination-option

21. Brian Glyn Williams, *Predators: The CIA's Drone War on al Qaeda* (Washington, DC: Potomac Books, 2013), 24. First quote is cited as Associated Press, "Officials: U.S. Missed Chance to Kill Bin Laden," June 23, 2003; second quote is cited as Scott Shane, "CIA to Expand Use of Drones in Pakistan," *New York Times*, December 4, 2009, A1.

22. Williams, *Predators*, 37.

23. Calhoun, *We Kill because We Can*, 37.

24. Falk, "Why Drones Are More Dangerous Than Nuclear Weapons," 43.

25. Donald R. Hickey, *The War of 1812: A Forgotten Conflict* (Champaign: University of Illinois Press, 2012), 304, 305.

26. Davis graduated from West Point in 1828, and Johnston graduated one year later. Davis soon left the Army but returned to service as a colonel in the Mexican War and served as the U.S. secretary of war in the Pierce administration. Johnston remained in the Army until the start of the Civil War, giving him far more military experience, but Davis consistently acted as though they remained on the parade ground at West Point.

27. Quoted in James M. McPherson, *Tried by War: Lincoln as Commander in Chief* (New York: Penguin, 2008), 66.

28. Mark Clodfelter, *The Limits of Airpower: The Bombing of North Vietnam* (Lincoln: University of Nebraska Press, 1989), 122–124; H. R. McMaster, *Dereliction of Duty: Lyndon Johnson, Robert McNamara, the Joint Chiefs of Staff, and the Lies That Led to Vietnam* (New York: HarperCollins, 1997), 208, 209.

29. William Wagner, *Lightning Bugs and Other Reconnaissance Drones* (Fallbrook, CA; Aero Publishers, 1982), 201–206; William Wagner and William P. Sloan, *Fireflies and Other UAVs* (Leicester, UK: Midland, 1992), 11.

30. The stealthy characteristics of the Sentinel, which remain classified, probably helped it to avoid detection by Pakistani air defenses.

31. Patrick Rayermann, "Exploiting Commercial SATCOM: A Better Way," *Parameters* (Winter 2003–2004), 54, 55; "Satellite Bandwidth," GlobalSecurity.org, http://www.globalsecurity.org/space/systems/bandwidth.htm

32. Steve Coll, *Ghost Wars: The Secret History of the CIA, Afghanistan, and Bin Laden, from the Soviet Invasion to September 10, 2001* (New York: Penguin, 2005), 581.

33. David Cloud and Greg Jaffe, *The Fourth Star* (New York: Three Rivers Press, 2009), 155–160.

34. Matt J. Martin and Charles Sasser, *Predator: The Remote-Control Air War over Iraq and Afghanistan: A Pilot's Story* (Minneapolis, MN: Zenith Press, 2010), 130–133; 219–224.

35. Ibid., 11.

36. Abraham D. Sofaer, "Responses to Terrorism: Targeted Killing Is a Necessary Option," *San Francisco Chronicle*, March 26, 2009, http://www.sfgate.com/opinion/openforum/article/Responses-to-Terrorism-Targeted-killing-is-a-2775845.php.

37. Lynn E. Davis, Michael McNerney, and Michael D. Greenberg, *Clarifying the Rules for Targeted Killing* (Santa Monica, CA: RAND Corporation, 2016), 9, 10.

38. Chris Woods, *Sudden Justice: America's Secret Drone Wars* (New York: Oxford University Press, 2015), 160.

39. Jo Becker and Scott Shane, "Secret 'Kill List' Proves a Test of Obama's Principles and Will," *New York Times*, May 29, 2012, http://www.nytimes.com/2012/05/29/world/obamas-leadership-in-war-on-al-qaeda.html

40. American Civil Liberties Union, "Targeted Killing," 2017, https://www.aclu.org/issues/national-security/targeted-killing

41. Sanger, *Confront and Conceal*, 260.

42. Woods, *Sudden Justice*, 202.

43. John P. Sullins, "Aspects of Telerobotic Systems," in Gerhard Dabringer, ed., *Ethical and Legal Aspects of Unmanned Systems—Interviews* (Wien, Germany: Institut für Religion und Frieden, 2010), 162.

44. Falk, "Why Drones Are More Dangerous Than Nuclear Weapons," 45.

45. Mary Ellen O'Connell, "Unlawful Killing with Combat Drones: A Case Study of Pakistan, 2004–2009," Legal Studies Research Paper No. 09–43 (South Bend, IN: Notre Dame Law School, 2009). Quotation is from the Abstract.

46. Human Rights Watch, "Letter to Obama on Targeted Killings and Drones," December 7, 2010, www.hrw.org/news/2010/12/07/letter-obama-targeted-killings

CHAPTER 6

1. Brian Orend, *The Morality of War*, 2nd ed. (Peterborough: Broadview Press, 2013), 92.

2. Avery Plaw, Matthew S. Fricker, and Carlos R. Colon, *The Drone Debate: A Primer on the U.S. Use of Unmanned Aircraft Outside Conventional Battlefields* (Lanham, MD: Rowman & Littlefield, 2016), 218.

3. Ibid., 168; Brian Glyn Williams, *Predators: The CIA's Drone War on al Qaeda* (Washington, DC: Potomac Books, 2013), 182–189.

4. Steven Pinker, *The Better Angels of Our Nature: Why Violence Has Declined* (New York: Penguin, 2012), 6–11.

5. Samuel Griffith, *Sun Tzu: The Art of War* (Oxford: Oxford University Press, 1971), 76.

6. Ibid., 84.

7. Carl von Clausewitz, *On War*, Michael Howard and Peter Paret, ed. and trans. (Princeton, NJ: Princeton University Press, 1976), 89.

8. Orend, *The Morality of War*, 125.

9. Tonkin Gulf Resolution, Public Law 88–408, 88th Congress, August 7, 1964.

10. Dana Wegner, "New Interpretations of How the USS Maine Was Lost," in Edward J. Marolda, ed., *Theodore Roosevelt, the U.S. Navy, and the Spanish-American War* (New York: Palgrave, 2001), 8-13.

11. Peter L. Bergen and Daniel Rothenberg, *Drone Wars: Transforming Conflict, Law, and Policy* (New York: Cambridge University Press, 2015), 288.

12. Plaw, Fricker, and Colon, *The Drone Debate*, 184–187.

13. Orend, *The Morality of War*, 113.

14. Bergen and Rothenberg, *Drone Wars*, 287.

15. Benjamin S. Lambeth, *The Unseen War: Allied Air Power and the Takedown of Saddam Hussein* (Annapolis, MD: Naval Institute Press, 2013), 294; Michael DeLong, *Inside CentCom: The Unvarnished Truth about the Wars in Afghanistan and Iraq* (Washington, DC: Regnery, 2004), 129.

16. Benjamin S. Lambeth, *NATO's Air War for Kosovo* (Santa Monica, CA: RAND Corporation, 2001), 144–147.

17. U.S. officials estimated 70 to 90 percent of civilians fled before the battle. See Dexter Filkins and James Glanz, "With Airpower and Armor, Troops Enter Rebel-Held City," *New York Times*, November 8, 2004, http://www.nytimes.com/2004/11/08/international/with-airpower-and-armor-troops-enter-rebelheld-city.html

18. U.S. War Department, Army War Plans Division, "AWPD-1" (Maxwell Air Force Base, AL: Air Force Historical Research Agency), August 12, 1941, 2.

19. Robert B. Strassler and Richard Crawley, *The Landmark Thucydides: A Comprehensive Guide to the Peloponnesian War* (New York: Free Press, 2008), 5.89.

20. Ann Rogers and John Hill, *Unmanned: Drone Warfare and Global Security* (London: Pluto Press, 2014), 110.

21. Lynn E. Davis, Michael McNerney, and Michael D. Greenberg, "Clarifying the Rules for Targeted Killing" (Santa Monica, CA: RAND Corporation, 2016), 5–9.

22. Peter Asaro, "Military Robots and Just War Theory," in Gerhard Dabringer, ed., *Ethical and Legal Aspects of Unmanned Systems—Interviews* (Wien, Germany: Institut für Religion und Frieden, 2010), 115.

23. Plaw, Fricker, and Colon, *The Drone Debate*, 179–181.

24. William T. Sherman to James M. Calhoun, E.E. Rawson, and S.C. Wells, *Official Records of the War of the Rebellion*, Series 1, Vol. 39, Part 2, (Washington, DC: Government Printing Office, 1881–1901), (September 12, 1864), 418.

25. James M. McPherson, *Battle Cry of Freedom: The Civil War Era* (New York: Oxford University Press, 2003), 774–776, 825–828.

26. Richard Miles, *Carthage Must Be Destroyed: The Rise and Fall of an Ancient Civilization* (London: Allen Lane, 2008), 338–348.

27. Ibid., 343–351.

28. Armin Krishnan, "Ethical and Legal Challenges," in Dabringer, ed., *Ethical and Legal Aspects of Unmanned Systems* (Wien, Germany: Institutfür Religion und Frieden, 2010), 68.

29. Ronald C. Arkin, *Governing Lethal Behavior in Autonomous Robots* (Boca Raton, FL: Taylor & Francis Group, 2009), 211.

30. Medea Benjamin, "The Grim Toll Drones Take on Innocent Lives," in Marjorie Cohn, ed., *Drones and Targeted Killing: Legal, Moral, and Geopolitical Issues.* (Northampton, MA: Olive Branch Press, 2015), 97.

31. Phyllis Bennis, "Drones and Assassination in the US's Permanent War," in Cohn, ed., *Drones and Targeted Killing* (Northampton, MA: Olive Branch Press, 2014), 57.

32. Michael Walzer, "The Triumph of Just War Theory (and the Dangers of Success)," in *Arguing About War* (New Haven, CT: Yale University Press, 2015), 16.

33. Grégoire Chamayou, *A Theory of the Drone*, Janet Lloyd, trans. (New York: The New Press, 2015), 213.

34. John Canning, "You've Just Been Disarmed. Have a Nice Day!" *IEEE Society and Technology Magazine* 28, no. 1 (Spring 2009): 13–15.

35. Colin Allen, "Morality and Artificial Intelligence," in Dabringer, ed., *Ethical and Legal Aspects of Unmanned Systems* (Wien, Germany: Institutfür Religion und Frieden, 2010), 24–27.

36. John Kaag and Sarah Kreps, *Drone Warfare* (Malden, MA: Polity Press, 2014), 115.

CHAPTER 7

1. Center for Strategic and International Studies, "Net Losses: Estimating the Global Cost of Cybercrime," June 2014, http://www.mcafee.com/us/resources/reports/rp-economic-impact-cybercrime2.pdf

2. In so doing, the Russians inadvertently used the same logic that Adolf Hitler cited when he announced a determination to annex the Sudetenland in 1939. Regrettably, the international response (protest, but take little specific action of consequence) was also approximately the same in both situations.

3. Senator Sam Nunn, quoted in Charles L. Fox and Dino A. Lorenzini, "How Much Is Not Enough? The Non-Nuclear Air Battle in NATO's Central Region," *Naval War College Review* 33 (March–April 1980): 68.

4. Harold C. Hutchison, "Ambitious Russian UAV Programs," *Strategy Page*, November 26, 2005, http://www.strategypage.com/htmw/htairfo/articles/2005 1126.aspx

5. Avery Plaw, Matthew S. Fricker, and Carlos R. Colon, *The Drone Debate: A Primer on the U.S. Use of Unmanned Aircraft Outside Conventional Battlefields* (Lanham, MD: Rowman & Littlefield, 2016), 285.

6. Mandiant Corporation, "APT1: Exposing One of China's Cyber Espionage Units," February 18, 2013, www.fireeye.com/content/dam/fireeye-www/ services/pdfs/mandiant-apt1-report.pdf; David E. Sanger, David Barboza, and Nicole Perlroth, "Chinese Army Unit Is Seen as Tied to Hacking against U.S.," *New York Times*, February 18, 2013.

7. P. W. Singer, *Wired for War: The Robotics Revolution and Conflict in the 21st Century* (New York: Penguin, 2009), 246.

8. Plaw, Fricker, and Colon, *The Drone Debate*, 285.

9. Dan Williams, "Israel Plans Laser Interceptor 'Iron Beam' for Short-Range Rockets," *Reuters*, January 19, 2014, https://www.reuters.com/article/us-arms-israel-interceptor-idUSBREA0I06M20140119; Tamir Eshel, "RAFAEL Develops a New High Energy Laser Weapon," *Defense Update*, January 19, 2014, http://defense-update.com/20140119_rafael-develops-new-high-energy-laser-weapon.html

10. Plaw, Fricker, and Colon, *The Drone Debate*, 311.

11. Barbara Opall-Rome, "Israel's Heavy-Hauling UAVs Are Ready for Battle," *Defense News*, January 25, 2010, http://www.defensenews.com/story.php?i=446 9090&c=MID&s=AIR

12. Paul J. Springer, *Military Robots and Drones* (Santa Barbara, CA: ABC-CLIO, 2013), 99–102.

13. John F. Gantz estimates that up to 90 percent of the world's computers run illegal copies of Microsoft Windows. This problem is particularly true in China and sub-Saharan Africa, making home computer users in those locations incredibly vulnerable to simple computer malware that has been long obsolete in regions where computer updates are the norm. See also Martin C. Libicki, *Cyberspace in Peace and War* (Annapolis, MD: Naval Institute Press, 2016), 11.

14. Reportedly, the Iranians used hot glue guns to fill the USB ports of all of their Natanz computers in the aftermath of this attack, a crude but effective way of making sure unauthorized devices will not be connected to the network. The very fact that the Iranian computers at Natanz had USB ports at all illustrates one of the dangers of using COTS technology—it can create vulnerabilities from its very design that are not easily countered.

15. Wilson W.S. Wong, *Emerging Military Technologies: A Guide to the Issues* (Santa Monica, CA: Praeger, 2013), 88.

16. Martin C. Libicki, *Cyberspace in Peace and War* (Annapolis, MD: Naval Institute Press, 2016), 14–19.

17. Thomas Rid, *Cyber War Will Not Take Place* (Oxford: Oxford University Press, 2013), 55.

CHAPTER 8

1. Micah Zenko, "100% Right 0% of the Time: Why the U.S. Military Can't Predict the Next War," *Foreign Policy*, October 16, 2012, http://foreignpolicy .com/2012/10/16/100-right-0-of-the-time/

2. For an outstanding primer on likely near-term future developments in robotics, see Illah Reza Nourbakhsh, *Robot Futures* (Cambridge: MIT Press, 2013), 27–48.

3. For more on Capek, see Peter Kussi and Arthur Miller, *Toward the Radical Center: A Karel Capek Reader* (North Haven, CT: Catbird Press, 1990).

4. Wilson W.S. Wong, *Emerging Military Technologies: A Guide to the Issues* (Santa Barbara, CA: Praeger, 2013).

5. Isaac Asimov, "Runaround," *I, Robot* (New York: Doubleday, 1950), 40.

6. Isaac Asimov, "The Evitable Conflict," *Robots and Empire* (New York: Doubleday, 1985), 186.

7. SRI International, "The Centibots Project," http://www.ai.sri.com/centibots/

8. Tim Arango and Michael R. Gordon, "Iraqi Prime Minister Arrives in Mosul to Declare Victory Over ISIS," *New York Times*, July 10, 2017, A1.

9. Lambér Royakkers and Rinie van Est, *Just Ordinary Robots: Automation from Love to War* (Boca Raton, FL: CRC Press, 2016), 284.

10. Mitch Utterback, "How ISIS Is Turning Commercial Drones into Weapons in the Battle for Mosul," *Fox News*, January 25, 2017, http://www.foxnews.com/tech/2017/01/25/how-isis-is-turning-commercial-drones-into-weapons-in-battle-for-mosul.html

11. The Heritage Foundation, "2017 Index of Military Strength," http://index.heritage.org/military/2017/assessments/us-military-power/u-s-army/

12. Kelly Beaucar Vlahos, "Handful of Lawmakers Send Their Kids to War," *Fox News*, March 28, 2003, http://www.foxnews.com/story/2003/03/28/handful-lawmakers-send-their-kids-to-wa-896091790.html; Amanda Terkel, "Veterans in New Congress Fewest since World War II," *Huffington Post*, November 17, 2012, https://www.huffingtonpost.com/2012/11/17/veterans-congress-fewest_n_2144852.html. In 1975, 306 members of Congress and 73 senators were veterans; by 2003, that number had dropped to 114 congressional members and 35 senators.

13. Mona Chalabi, "What Percentage of Americans Have Served in the Military?" *FiveThirtyEight*, March 19, 2015, https://fivethirtyeight.com/features/what-percentage-of-americans-have-served-in-the-military/.

14. In 2000, General Hugh Shelton referred to the "Dover Test," an informal measure of whether the U.S. population would tolerate the return of flag-draped coffins from a war zone to the Dover air force base. See Henry H. Shelton, "National Security and the Intersection of Force and Diplomacy," Remarks to the ARCO Forum (Kennedy School of Government, Harvard University, January 19, 2000).

15. Medea Benjamin, "The Grim Toll Drones Take on Innocent Lives," in Marjorie Cohn, ed., *Drones and Targeted Killing: Legal, Moral, and Geopolitical Issues* (Northampton, MA: Olive Branch Press, 2015), 97. Benjamin was referring to the killing of Abduhrahman al-Awlaki, whose case is discussed in Chapter 4 of this work.

16. Barry Strauss, *Masters of Command: Alexander, Hannibal, Caesar, and the Genius of Leadership* (New York: Simon & Schuster, 2012), 160–163.

17. Dee Brown, *The Galvanized Yankees* (Lincoln: University of Nebraska Press, 1963), 211–216.

18. "Hitachi's New System Identifies Based on Face, Attire, Age, Sex, and More," *FindBiometrics*, March 27, 2017, http://www.findbiometrics.com/hitachi-system-face-attire-age-403275/

19. Mark Thompson, "The Curse of 'Friendly Fire,'" *Time*, June 10, 2014, http://www.time.com/2854306/the-curse-of-friendly-fire/

20. Pamela Hess, "The Patriot's Fratricide Record," *UPI*, April 24, 2003, http://www.upi.com/Feature-The-Patriots-fratricide-record/63991051224638/

21. Noah Shachtman, "Robot Cannon Kills 9, Wounds 14," *Wired*, October 18, 2007, http://www.wired.com/2007/robot-cannon.ki/

22. Lynn E. Davis, Michael McNerney, and Michael D. Greenberg, "Clarifying the Rules for Targeted Killing" (Santa Monica, CA: RAND Corporation, 2016), 18.

23. Sanger, *Confront and Conceal*, 261.

24. Falk, "Why Drones Are More Dangerous Than Nuclear Weapons," 45.

25. Chamayou, *A Theory of the Drone*, 202.

26. Calhoun, *We Kill because We Can*, xvi.

27. Davis, McNerney, and Greenberg, *Clarifying the Rules for Targeted Killing*, 16, 17.

28. M. Schulzke, "Autonomous Weapons and Distributed Responsibility," *Philosophy and Technology* 26(2): 215.

Bibliography

Abé, Nicola. "Dreams in Infrared: The Woe of an American Drone Operator." *Spiegel Online*, December 14, 2012, https://www.sott.net/article/254875-Dreams-in-infrared-The-woes-of-an-American-drone-operator

Ackerman, Evan. "We Should Not Ban 'Killer Robots,' and Here's Why." *IEEE Spectrum*, July 29, 2015, http://spectrum.ieee.org/automaton/robotics/artificial-intelligence/we-should-not-ban-killer-robots

Ackerman, Spencer, Ed Pilkington, Ben Jacobs, and Julian Borger. "Syria Missile Strikes: US Launches First Direct Military Action against Assad." *The Guardian*, April 7, 2017, https://www.theguardian.com/world/2017/apr/06/trump-syria-missiles-assad-chemical-weapons

Al-Awlaki, Nasser. "The Drone That Killed My Grandson." *The New York Times*, July 17, 2013, http://www.nytimes.com/2013/07/18/opinion/the-drone-that-killed-my-grandson.html

American Civil Liberties Union. "Targeted Killing." https://www.aclu.org/issues/national-security/targeted-killing

Arango, Tim, and Michael R. Gordon. "Iraqi Prime Minister Arrives in Mosul to Declare Victory Over ISIS." *The New York Times*, July 9, 2017, https://www.nytimes.com/2017/07/09/world/middleeast/mosul-isis-liberated.html

Arkin, Ronald C. *Governing Lethal Behavior in Autonomous Robots*. Boca Raton, FL: Taylor & Francis Group, 2009.

Armitage, Michael. *Unmanned Aircraft*. London: Brassey's Defence Publishers, 1988.

Asimov, Isaac. *I, Robot*. New York: Doubleday, 1950.

Asimov, Isaac. *Robots and Empire*. New York: Doubleday, 1985.

Associated Press. "US-Born Radical Cleric Added to Terror Blacklist." *Fox News*, July 16, 2010, http://www.foxnews.com/us/2010/07/16/born-radical-cleric-added-terror-blacklist.html

Baggesen, Arne. "Design and Operational Aspects of Autonomous Unmanned Combat Aerial Vehicles." Monterey, CA: Naval Postgraduate School, 2005.

Bazan, Elizabeth B. "Assassination Ban and E.O. 12333: A Brief Summary." Washington, DC: Congressional Research Service, January 4, 2002.

Becker, Jo, and Scott Shane. "Secret 'Kill List' Proves a Test of Obama's Principles and Will." *The New York Times*, May 29, 2012, http://www.nytimes.com/2012/05/29/world/obamas-leadership-in-war-on-al-qaeda.html

Bergen, Peter L., and Daniel Rothenberg. *Drone Wars: Transforming Conflict, Law, and Policy*. New York: Cambridge University Press, 2015.

Bergström, Christer. *Barbarossa—The Air Battle: July–December 1941*. Crowborough, UK: Classic Publications, 2003.

Berkowitz, Bruce. "Is Assassination an Option?" *Hoover Digest*, January 30, 2002, www.hoover.org/research/assassination-option

Bishop, M.C., and J.C.N. Coulston. *Roman Military Equipment from the Punic Wars to the Fall of Rome*. Oxford: Oxford University Press, 2006.

Blair, D., and A. Spillius. "Iran Shows Off Captured U.S. Drone." *The Telegraph*, December 8, 2011, http://www.telegraph.co.uk/news/worldnews/middleeast/iran/8944248/Iran-shows-off-captured-US-drone.html

Boone, Jeb, and Greg Miller. "U.S. Drone Strike in Yemen Is First since 2002." *Washington Post*, May 5, 2011, https://www.washingtonpost.com/world/middle-east/yemeni-official-us-drone-strike-kills-2-al-qaeda-operatives/2011/05/05/AF7HrzxF_story.html?utm_term=.e7c492bf16d9

Boston Dynamics. "BigDog: The First Advanced Rough-Terrain Robot." https://www.bostondynamics.com/bigdog

Bowden, Mark. "The Killing Machines: How to Think about Drones." *The Atlantic*, September 2013, https://www.theatlantic.com/magazine/archive/2013/09/the-killing-machines-how-to-think-about-drones/309434/

Bowers, Ray L. *Tactical Airlift*. Washington, DC: Office of Air Force History, 1999.

Brissett, Wilson. "ISR Explosion." *Air Force Magazine* (July 2017): 48–55.

Brodie, Bernard. *Strategy in the Missile Age*. Santa Monica, CA: RAND Corporation, 1959.

Brooks, Rodney. "Fast, Cheap, and Out of Control: A Robot Invasion of the Solar System." *Journal of the British Interplanetary Society* 42 (1989): 478–485.

Brooks, Rodney. *Flesh and Machines: How Robots Will Change US*. New York: Pantheon Books, 2002.

Brown, Dee. *The Galvanized Yankees*. Lincoln: University of Nebraska Press, 1963.

Brown, Robbie, and Kim Severson. "2nd American in Strike Waged Qaeda Media War." *The New York Times*, September 30, 2011, A1.

Browning, Christopher R. *Ordinary Men: Reserve Police Battalion 101 and the Final Solution in Poland*. New York: HarperCollins, 1992.

Burnette, Heyward. "AFRL Incorporates Solar Cell Technology into Small Unmanned Aircraft Systems." *SUAS News*, November 17, 2012, https://www.suasnews.com/2012/11/afrl-incorporates-solar-cell-technology-into-small-unmanned-aircraft-systems/

Bush, George W. *Decision Points*. New York: Crown Publishers, 2010.

Butt, David S. "U.S.-Born Cleric Was Target of Yemen Drone Strike." *Los Angeles Times*, May 7, 2011, http://articles.latimes.com/2011/may/07/world/la-fg-yemen-drones-20110507

Calhoun, Laurie. *We Kill Because We Can: From Soldiering to Assassination in the Drone Age*. London: Zed Books, 2015.

Canning, John. "You've Just Been Disarmed. Have a Nice Day!" *IEEE Society and Technology Magazine* 28, no. 1 (Spring 2009): 13–15.

Capaccio, Anthony. "F-35 Program Costs Jump to $406.5 Billion in Latest Estimate." *Bloomberg*, July 10, 2017, https://www.bloomberg.com/news/articles/2017-07-10/f-35-program-costs-jump-to-406-billion-in-new-pentagon-estimate

Center for Strategic and International Studies. "Net Losses: Estimating the Global Cost of Cybercrime." Center for Strategic and International Studies, June 2014, http://www.mcafee.com/us/resources/reports/rp-economic-impact-cybercrime2.pdf

Chalabi, Mona. "What Percentage of Americans Have Served in the Military?" *FiveThirtyEight*, March 19, 2015, https://fivethirtyeight.com/features/what-percentage-of-americans-have-served-in-the-military/

Chamayou, Grégoire. *A Theory of the Drone*. Janet Lloyd, trans. New York: The New Press, 2015.

Christopher, Russell. "Imminence in Justified Targeted Killing," in Clair Oakes Finkelstein, Jens David Ohlin, and Andrew Altman, eds. *Targeted Killing*. Oxford, UK: Oxford University Press, 2012, 253–284.

Clark, Colin, and Sydney J. Freedberg, Jr. "A-10: Close Air Support Wonder Weapon or Boneyard Bound?" *Breaking Defense*, December 19, 2013, http://breakingdefense.com/2013/12/a-10-close-air-support-wonder-weapon-or-boneyard-bound/

Clark, Richard M. "Uninhabited Combat Aerial Vehicles: Airpower by the People, for the People, but Not with the People." CADRE Paper No. 8, Maxwell Air Force Base, AL: Air University Press, 2000.

Clausewitz, Carl von. *On War*. Michael Howard and Peter Paret, ed., trans. Princeton, NJ: Princeton University Press, 1976.

Clodfelter, Mark. *The Limits of Airpower: The American Bombing of North Vietnam*. Lincoln: University of Nebraska Press, 1989.

Cloud, David, and Greg Jaffe. *The Fourth Star*. New York: Three Rivers Press, 2009.

Cohen, William J., and Ken Zemach. "The MARCbot: The Army Program That Revolutionized Robotics for Patrol Warfighters." *IQT Quarterly* 3 (Summer 2011): 15–17.

Cohn, Marjorie, ed. *Drones and Targeted Killing: Legal, Moral, and Geopolitical Issues*. Northampton, MA: Olive Branch Press, 2015.

Coll, Steve. *Ghost Wars: The Secret History of the CIA, Afghanistan, and Bin Laden, from the Soviet Invasion to September 10, 2001*. New York: Penguin, 2005.

Congressional Budget Office. "Policy Options for Unmanned Aircraft Systems." Washington, DC: Congressional Budget Office, 2011.

Cowan, Ross. "Equipment." *Roman Legionary: 58 BC–AD 69*. Oxford: Osprey Publishing, 2003.

Dabringer, Gerhard. *Ethical and Legal Aspects of Unmanned Systems—Interviews*. Wien, Germany: Institutfür Religion und Frieden, 2010.

Daly, Mark, ed. *Jane's Unmanned Aerial Vehicles and Targets*. Alexandria, VA: Jane's Information Group, 2009.

Davis, Lynn E., Michael McNerney, and Michael D. Greenberg. "Clarifying the Rules for Targeted Killing." Santa Monica, CA: RAND Corporation, 2016.

De Luce, Dan, Robbie Gramer, and Jana Winter. "John Walker Lindh, Detainee #001 in the Global War on Terror, Will Go Free in Two Years. What Then?"

Foreign Policy, January 11, 2013, http://foreignpolicy.com/2017/06/23/
john-walker-lindh-detainee-001-in-the-global-war-on-terror-will-go-free-
in-two-years-what-then/

DeLong, Michael. *Inside CentCom: The Unvarnished Truth about the Wars in Afghanistan and Iraq*. Washington, DC: Regnery, 2004.

Editorial Board. "Libya and the War Powers Act: The Law Does Apply to the
NATO Campaign, but That Is No Excuse to End It Prematurely." *The New
York Times*, June 17, 2011, A34.

Eisenhower, John S.D. *Agent of Destiny: The Life and Times of General Winfield Scott*.
Norman: University of Oklahoma Press, 1999.

Eshel, Tamir. "RAFAEL Develops a New High Energy Laser Weapon." *Defense
Update*, January 19, 2014, http://defense-update.com/20140119_rafael-
develops-new-high-energy-laser-weapon.html

Farrier, John. "America's Monument to Its Most Infamous Traitor, Benedict Arnold."
Neatorama, January 1, 2014, http://www.neatorama.com/2014/01/01/
Americas-Monument-to-Its-Most-Infamous-Traitor-Benedict-Arnold/

Federation of American Scientists. "Run Silent, Run Deep." December 8, 1998,
https://fas.org/man/dod-101/sys/ship/deep.htm

Ferrill, Arthur. *The Origins of War: From the Stone Age to Alexander the Great*. London: Thames and Hudson, 1986.

Filkins, Dexter, and James Glanz. "With Airpower and Armor, Troops Enter Rebel-
Held City." *The New York Times*, November 8, 2004, http://www.nytimes
.com/2004/11/08/international/with-airpower-and-armor-troops-enter-
rebelheld-city.html

Finn, Anthony, and Steve Scheding. *Developments and Challenges for Autonomous
Unmanned Vehicles: A Compendium*. Heidelberg, Germany: Springer, 2010.

Finn, Tom, and Noah Browning. "An American Teenager in Yemen: Paying for the
Sins of His Father?" *Time*, October 27, 2011, http://content.time.com/time/
world/article/0,8599,2097899,00.html

Fitzgerald, Mary C. "Marshal Ogarkov and the New Revolution in Soviet Military Affairs." Washington, DC: Center for Naval Analyses, Office of Naval
Research, January 14, 1987.

Friedersdorf, Conor. "How Team Obama Justifies the Killing of a 16-Year-Old
American." *The Atlantic*, October 24, 2012, https://www.theatlantic
.com/politics/archive/2012/10/how-team-obama-justifies-the-killing-of-
a-16-year-old-american/264028/

Future of Life Institute. "Autonomous Weapons: An Open Letter from AI &
Robotics Researchers." International Joint Conference on Artificial Intelligence, Buenos Aires, Argentina, July 28, 2015, https://futureoflife.org/
open-letter-autonomous-weapons#

Gantz, John F., et al. "The Dangerous World of Counterfeit and Pirated Software."
Microsoft White Paper, 2013, http://news.microsoft.com/download/
presskits/antipiracy/docs/IDC030513.pdf

Gates, Robert M. *Duty: Memoirs of a Secretary at War*. New York: Alfred A. Knopf,
2014.

Gertler, Jeremiah. "U.S. Unmanned Systems." Washington, DC: Congressional
Research Service, 2012.

Goldhagen, Daniel Jonah. *Hitler's Willing Executioners: Ordinary Germans and the Holocaust*. New York: Vintage, 1996.

Gorman, Siobhan, Yochi J. Dreazen, and August Cole. "Insurgents Hack U.S. Drones." *Wall Street Journal*, December 17, 2009, https://www.wsj.com/articles/SB126102247889095011

Greeley, Horace. *The American Conflict: A History of the Great Rebellion in the United States of America, 1860–1864*, Vol. 1. New York: O.D. Case, 1864.

Griffith, Samuel. *Sun Tzu: The Art of War*. Oxford: Oxford University Press, 1971.

Griswold, Mary E. "Spectrum Management: Key to the Future of Unmanned Aircraft Systems?" Maxwell Paper No. 44, Maxwell Air Force Base, AL: Air University Press, 2008.

Grossman, Dave. *On Killing: The Psychological Cost of Learning to Kill in War and Society*. New York: Back Bay Books, 1995.

Groves, Leslie M. *Now It Can Be Told: The Story of the Manhattan Project*. New York: Harper, 1962.

Guetlein, Mike. *Lethal Autonomous Weapons—Ethical and Doctrinal Implications*. Newport, RI: Naval War College, 2005.

Hambling, David. "Russian Drones Attack with Grenade Weapons." *Scout Warrior*, July 18, 2017, http://scout.com/military/warrior/Article/Small-Russian-Drones-Do-Massive-Damage-WIth-Grenade-Weapons-103103172

Harford, Tim. *Adapt: Why Success Always Starts with Failure*. New York: Picador, 2011.

Haulman, Daniel. "U.S. Unmanned Aerial Vehicles in Combat, 1991–2003." Maxwell Air Force Base, AL: U.S. Air Force Historical Research Agency, 2003.

The Heritage Foundation. "2017 Index of U.S. Military Strength," 2017, http://index.heritage.org/military/2017/assessments/us-military-power/u-s-army/

Hess, Pamela. "The Patriot's Fratricide Record." *UPI*, April 24, 2003, http://www.upi.com/Feature-The-Patriots-fratricide-record/63991051224638/

Hickey, Donald R. *The War of 1812: A Forgotten Conflict*. Champaign: University of Illinois Press, 2012.

"Hitachi's New System Identifies Based on Face, Attire, Age, Sex, and More." *FindBiometrics*, March 27, 2017, http://www.findbiometrics.com/hitachi-system-face-attire-age-403275/

Holland, Tim. *Persian Fire: The First World Empire and the Battle for the West*. London: Abacus Press, 2006.

Hughes, Thomas. *Over Lord: General Pete Quesada and the Triumph of Tactical Airpower in World War II*. New York: Free Press, 2002.

Human Rights Watch. "Letter to Obama on Targeted Killings and Drones." December 7, 2010, www.hrw.org/news/2010/12/07/letter-obama-targeted-killings

Hume, David B. "Integration of Weaponized Unmanned Aircraft into the Air-to-Ground System." Maxwell Paper No. 41, Maxwell Air Force Base, AL: Air War College, 2007.

Hutchison, Harold C. "Ambitious Russian UAV Programs." *Strategy Page*, November 26, 2005, http://www.strategypage.com/htmw/htairfo/articles/20051126.aspx.

International Atomic Energy Agency. "Treaty on the Non-Proliferation of Nuclear Weapons." 1970.

Jelinek, Pauline. "Pentagon: Insurgents Intercepted UAV Videos." *Army Times*, December 17, 2009, http://www.armytimes.com/news/2009/12/ap_uav_ insurgents_hacked_121709/

Johnson-Freese, Joan. *Heavenly Ambitions: America's Quest to Dominate Space*. Philadelphia: University of Pennsylvania Press, 2009.

"Justice Department Memo Reveals Legal Case for Drone Strikes on Americans." *NBC News*, February 4, 2013, http://investigations.nbcnews.com/_news/ 2013/02/04/16843014-justice-department-memo-reveals-legal-case-for- drone-strikes-on-americans

Kaag, John, and Sarah Kreps. *Drone Warfare*. Malden, MA: Polity Press, 2014.

Kahn, Herman. *On Thermonuclear War*. New York: Free Press, 1960.

Kahn, Paul. "The Paradox of Riskless War." *Philosophy and Public Policy Quarterly* 22, no. 3 (Summer 2002): 2–8.

Kamen, Al. "Drone Pilots to Get Medals?" *Washington Post*, September 7, 2012, https://www.washingtonpost.com/blogs/in-the-loop/post/drone-pilots- to-get-medals/2012/07/09/gJQAF2PhYW_blog.html?utm_term=.36d9 dead4bef

Kane, Tim. "Global U.S. Troop Deployment, 1950–2003." The Heritage Center for Data Analysis, October 27, 2004, http://www.heritage.org/defense/ report/global-us-troop-deployment-1950-2003

Keeley, Lawrence H. *War before Civilization: The Myth of the Peaceful Savage*. Oxford: Oxford University Press, 1996.

"Kellogg-Briand Pact." Avalon Project. Yale Law School, Lillian Goldman Law Library, 1928, http://avalon.law.yale.edu/20th_century/kbpact.asp

Khan, Sameen, and Salman Masood. "Suicide Bombing Targeting Pakistani Police Kills at Least 26." *The New York Times*, July 24, 2017, https://www.nytimes .com/2017/07/24/world/asia/pakistan-lahore-suicide-bombing.html

Koh, Harold. "Obama's ISIL Legal Rollout: Bungled, Clearly. But Illegal? Really?" *Just Security*, September 29, 2014, https://www.justsecurity.org/15692/ obamas-isil-legal-rollout-bungled-clearly-illegal-really/

Krentz, Peter. *The Battle of Marathon*. New Haven, CT: Yale University Press, 2011.

Kreps, Sarah E. *Drones: What Everyone Needs to Know*. New York: Oxford University Press, 2016.

Krishnan, Armin. *Killer Robots: Legality and Ethicality of Autonomous Weapons*. Burlington, VT: Ashgate, 2009.

Krivosheev, G. F. *Soviet Casualties and Combat Losses in the Twentieth Century*. London: Greenhill Books, 1997.

Kurzweil, Ray. *The Singularity Is Near: When Humans Transcend Biology*. New York: Penguin, 2005.

Kussi, Peter, ed. *Toward the Radical Center: A Karel Capek Reader*. Highland Park, NJ: Catbird Press, 1990.

Lambeth, Benjamin S. *NATO's Air War for Kosovo: A Strategic and Operational Assessment*. Santa Monica, CA: RAND Corporation, 2001.

Lambeth, Benjamin S. *The Unseen War: Allied Air Power and the Takedown of Saddam Hussein*. Annapolis, MD: Naval Institute Press, 2013.

"Leash the Dogs of War: South Africa Struggles in Vain to Ban Soldiers of Fortune." *The Economist*, March 19, 2015, https://www.economist.com/news/

middle-east-and-africa/21646809-south-africa-struggles-vain-ban-sol
    diers-fortune-leash-dogs-war
Lee, Barbara. "Why I Opposed the Resolution to Authorize Force." *SF Gate*,
    September 23, 2001, http://www.sfgate.com/opinion/article/Why-I-
    opposed-the-resolution-to-authorize-force-2876893.php
Lehner, Paul E. *Artificial Intelligence and National Defense: Opportunity and Challenge.*
    Blue Ridge Summit, PA: TAB Books, 1989.
Lendon, Brad. "ISIS May Have Saved the A-10." *CNN*, January 22, 2016, http://
    www.cnn.com/2016/01/21/politics/air-force-a-10-isis/index.html
Libicki, Marin C. *Cyberspace in Peace and War*. Annapolis, MD: Naval Institute
    Press, 2016.
Limbaugh, Rush. "The Miracle of Cochlear Implant Surgery." April 8, 2014,
    http://www.rushlimbaugh.com/daily/2014/04/08/the_miracle_of_coch
    lear_implant_surgery/
Lindh, Frank. "America's 'Detainee 001.'"*The Guardian*, July 10, 2011, https://
    www.theguardian.com/world/2011/jul/10/john-walker-lindh-american-
    taliban-father
Lowensohn, Josh. "Google Buys Boston Dynamics, Maker of Spectacular and
    Terrifying Robots." *The Verge*, December 14, 2013, https://www.theverge
    .com/2013/12/14/5209622/google-has-bought-robotics-company-boston-
    dynamics
Lunden, Ingrid. "SoftBank Is Buying Robotics Firms Boston Dynamics and
    Schaft from Alphabet." *Tech Crunch*, June 8, 2017, https://techcrunch
    .com/2017/06/08/softbank-is-buying-robotics-firm-boston-dynamics-
    and-schaft-from-alphabet/
Mandiant Corporation. "APT1: Exposing One of China's Cyber Espionage Units."
    February 18, 2013, www.fireeye.com/content/dam/fireeye-www/ser
    vices/pdfs/mandiant-apt1-report.pdf
Martin, Matt J., and Charles Strasser. *Predator: The Remote-Control Air War over Iraq
    and Afghanistan: A Pilot's Story*. Minneapolis, MN: Zenith Press, 2010.
Mazzetti, Mark, Charlie Savage, and Scott Shane. "A U.S. Citizen in America's
    Cross Hairs." *The New York Times*, March 10, 2013, A1.
McCaffery, James M. *Army of Manifest Destiny: The American Soldier in the Mexican
    War, 1846–1848*. New York: New York University Press, 1994.
McGarry, Brendan. "Army Not Interested in Taking A-10 Warthogs from Air Force."
    DoD Buzz, February 25, 2015, https://www.dodbuzz.com/2015/02/25/
    army-not-interested-in-taking-a-10-warthogs-from-air-force/
McMaster, H. R. *Dereliction of Duty: Lyndon Johnson, Robert McNamara, the Joint
    Chiefs of Staff, and the Lies that Led to Vietnam*. New York: HarperCollins, 1997.
McNeill, William H. *The Pursuit of Power: Technology, Armed Force, and Society since
    A.D. 1000*. Chicago: University of Chicago Press, 1982.
McPherson, James M. *Battle Cry of Freedom: The Civil War Era*. New York: Oxford
    University Press, 2003.
McPherson, James M. *Tried by War: Lincoln as Commander in Chief*. New York: Pen-
    guin Press, 2008.
Melzer, Nils. *Targeted Killing in International Law*. Oxford: Oxford University
    Press, 2008.

Miles, Richard. *Carthage Must Be Destroyed: The Rise and Fall of an Ancient Civilization*. London: Allen Lane, 2008.

Miller, W. Lee. "USMC Ground Robotics Current and Desired Future Capabilities." Defense Technical Information Center, 2009, www.dtic.mil/ndia/2009/groundrobot/miller.pdf

Millett, Allan R., Peter Maslowski, and William B. Feis. *For the Common Defense: A Military History of the United States from 1607 to 2012*. New York: Free Press, 2012.

Mindell, David A. *Our Robots, Ourselves: Robotics and the Myths of Autonomy*. New York: Viking, 2015.

Mizokami, Kyle. "Kaboom! Russian Drone with Thermite Grenade Blows Up a Billion Dollars of Ukrainian Ammo." *Popular Mechanics*, July 27, 2017, http://www.popularmechanics.com/military/weapons/news/a27511/russia-drone-thermite-grenade-ukraine-ammo/

Moore, Gordon E. "Cramming More Components onto Integrated Circuits." *Electronics* 38, no. 8 (April 19, 1965), 114–117.

Morelle, Rebecca. "Meet the Creatures That Live Beyond the Abyss." *BBC News*, January 22, 2010, http://news.bbc.co.uk/1/hi/8426132.stm

Mothana, Ibrahim. "How Drones Help Al Qaeda." *The New York Times*, June 14, 2012, A35.

Musa ibn Shakir, Muhammad ibn, Ahmad ibn Musa ibn Shakir, and Hasanibn Musa ibn Shakir. *The Book of Ingenious Devices*, Donald Routledge Hill, trans. Dordrecht, Netherlands: D. Reidel, 1979.

Nonami, Kenzo, Farid Kendoul, Satoshi Suzuki, Wei Wang, and Daisuke Nakazawa. *Autonomous Flying Robots: Unmanned Aerial Vehicles and Micro Aerial Vehicles*. New York: Springer, 2010.

Norton, David F. "The Military Adoption of Innovation." Monterey, CA: Naval Postgraduate School, 2007.

Nourbakhsh, Illah Reza. *Robot Futures*. Cambridge: MIT Press, 2013.

"Obama: Awlaki Death 'Major Blow' to Terror." *CBS News*, September 30, 2011, http://www.cbsnews.com/news/obama-awlaki-death-major-blow-to-terror/

"Obama: 'We Took Out' Pakistani Taliban Chief." *Reuters*, August 20, 2009, http://www.reuters.com/article/us-obama-pakistan-mehsud-idUSTRE57J5EC20090820

O'Connell, Mary Ellen. "Unlawful Killing with Combat Drones: A Case Study of Pakistan, 2004–2009." Legal Studies Research Paper No. 09–43, Notre Dame Law School, 2009.

Opall-Rome, Barbara. "Israel's Heavy-Hauling UAVs Are Ready for Battle." *Defense News*, January 25, 2010, http://www.defensenews.com/story.php?i=4469090&c=MID&s=AIR

Orend, Brian. *The Morality of War*, 2nd ed. Peterborough: Broadview Press, 2013.

"Over 700 Killed in 44 Drone Strikes in 2009." *Dawn* (Pakistan). January 2, 2010, https://www.dawn.com/news/958386

Panetta, Leon. *Worthy Fights: A Memoir of Leadership in War and Peace*. New York: Penguin Press, 2015.

Pape, Robert. *Bombing to Win: Air Power and Coercion in War*. Ithaca, NY: Cornell University Press, 1996.

Parker, Geoffrey. *The Military Revolution: Military Innovation and the Rise of the West, 1500–1800*. New York: Cambridge University Press, 1988.

Pawlyk, Oriana. "Report: A-10 Retirement Indefinitely Delayed." *Air Force Times*, January 13, 2016, https://www.airforcetimes.com/news/your-air-force/2016/01/13/report-a-10-retirement-indefinitely-delayed/

Perez, Evan. "Judge Dismisses Targeted-Killing Suit." *Wall Street Journal*, December 8, 2010, https://www.wsj.com/articles/SB10001424052748703296604576005391675065166

Perkowitz, Sidney. *Digital People: From Bionic Humans to Androids*. Washington, DC: Joseph Henry Press, 2004.

Pinker, Steven. *The Better Angels of Our Nature: Why Violence Has Declined*. New York: Penguin, 2012.

Plaw, Avery, Matthew S. Fricker, and Carlos R. Colon. *The Drone Debate: A Primer on the U.S. Use of Unmanned Aircraft Outside Conventional Battlefields*. Lanham, MD: Rowman & Littlefield, 2016.

Rayermann, Patrick. "Exploiting Commercial SATCOM: A Better Way." *Parameters*, Winter (2003–2004): 54–66.

Revill, Jo. " 'Remote' Surgery Turning Point." *The Guardian*, October 6, 2002, https://www.theguardian.com/society/2002/oct/06/health.medicineandhealth

Ricks, Thomas E. "Target Approval Cost Air Force Key Hits." *Washington Post*, November 18, 2001, http://www.washingtonpost.com/wp-dyn/articles/A46827-2001Nov17.html

Rid, Thomas. *Cyber War Will Not Take Place*. Oxford: Oxford University Press, 2013.

Riskin, Jessica. "The Defecating Duck, or, the Ambiguous Origins of Artificial Life." *Critical Inquiry* 29, no. 4 (Summer 2003): 599–633.

Rogers, Ann, and John Hill. *Unmanned: Drone Warfare and Global Security*. London: Pluto Press, 2014.

Rosen, Rebecca J. "Unimate: The Story of George Devol and the First Robotic Arm." *The Atlantic*, August 16, 2011, https://www.theatlantic.com/technology/archive/2011/08/unimate-the-story-of-george-devol-and-the-first-robotic-arm/243716/

Rosheim, Mark E. *Leonardo's Lost Robots*. New York: Springer, 2006.

Royakkers, Lambèr, and Rinie van Est. *Just Ordinary Robots: Automation from Love to War*. Boca Raton, FL: CRC Press, 2016.

Runciman, Steven. *The Fall of Constantinople, 1453*. Cambridge: Cambridge University Press, 1965.

Sanchez, Raf. " 'The Devil of Ramadi' Named America's Deadliest Sniper." *The Telegraph*, January 3, 2012, http://www.telegraph.co.uk/news/worldnews/northamerica/usa/8990552/The-Devil-of-Ramadi-named-Americas-deadliest-sniper.html

Sanger, David E. *Confront and Conceal: Obama's Secret Wars and Surprising Use of American Power*. New York: Crown Publishers, 2012.

Sanger, David E., David Barboza, and Nicole Perlroth. "Chinese Army Unit Is Seen as Tied to Hacking against U.S." *The New York Times*, February 19, 2013, A1.

Sarno, David. "Apple's Market Value Tops $500 Billion." *Los Angeles Times*, February 29, 2012, http://articles.latimes.com/2012/feb/29/business/la-fi-apple-value-20120301

"Satellite Bandwidth." GlobalSecurity.org. http://www.globalsecurity.org/space/systems/bandwidth.htm

Savage, Charlie. "Secret Memo Made Legal Case to Kill a Citizen." *The New York Times*, October 9, 2011, A1.

Savage, Charlie. "Was Trump's Syria Strike Illegal? Explaining Presidential War Powers." *The New York Times*, April 8, 2017, A11.

Savage, Charlie, and Mark Landler. "White House Defends Continuing U.S. Role in Libya Operation." *The New York Times*, June 16, 2011, A16.

Scahill, Jeremy. *Dirty Wars: The World Is a Battlefield*. New York: Nation Books, 2013.

Scharre, Paul. "Robotics on the Battlefield, Part I: Range, Persistence and Daring." Washington, DC: Center for a New American Security, May 2014.

Schulzke, M. "Autonomous Weapons and Distributed Responsibility." *Philosophy and Technology* 26 (2): 203–291.

Shachtman, Noah. "First Armed Robots on Patrol in Iraq." *Wired*, August 2, 2007, https://www.wired.com/2007/08/httpwwwnational

Shachtman, Noah. "Robot Cannon Kills 9, Wounds 14." *Wired*. October 18, 2007, https://www.wired.com/2007/10/robot-cannon-ki/

Shaheen, Kareem. "Assad Forces Carried Out Sarin Attack, Says French Intelligence." *The Guardian*, April 26, 2017, https://www.theguardian.com/world/2017/apr/26/syria-assad-forces-carried-out-sarin-attack-says-french-intelligence

Shane, Scott. "CIA to Expand Use of Drones in Pakistan." *The New York Times*, December 4, 2009, A1.

Sharkey, Noel. "America's Mindless Killer Robots Must Be Stopped." *The Guardian*, December 3, 2012, https://www.theguardian.com/commentisfree/2012/dec/03/mindless-killer-robots

Sharp, Jane. *Striving for Military Stability in Europe*. New York: Routledge, 2010.

Sheehan, Neil. *A Fiery Peace in a Cold War: Bernard Schriever and the Ultimate Weapon*. New York: Random House, 2009.

Shelton, Henry H. "National Security and the Intersection of Force and Diplomacy," Remarks to the ARCO Forum. Cambridge, MA: Kennedy School of Government, Harvard University, January 19, 2000.

Siddiqui, Sabrina. "Obama 'Surprised,' 'Upset' When Anwar Al-Awlaki's Teenage Son Was Killed by U.S. Drone Strike." *Huffington Post*, April 23, 2013, https://www.huffingtonpost.com/2013/04/23/obama-anwar-al-awlaki-son_n_3141688.html

Singer, P. W. "Caution: Children at War." *Parameters*, Winter (2001–2002): 40–56.

Singer, P. W. *Children at War*. Berkeley: University of California Press, 2009.

Singer, P. W. *Corporate Warriors: The Rise of the Privatized Military Industry*. Ithaca, NY: Cornell University Press, 2003.

Singer, P. W. *Wired for War: The Robotics Revolution and Conflict in the 21st Century*. New York: Penguin, 2009.

"60 Drone Hits Kill 14 Al-Qaeda Men, 687 Civilians." *News* (Pakistan), April 10, 2009, http://www.ikhwanweb.com/article.php?id=19870.

Smith, David. "South Africa's Ageing White Mercenaries Who Helped Turn Tide on Boko Haram." *The Guardian*, April 14, 2015, https://www.theguardian

.com/world/2015/apr/14/south-africas-ageing-white-mercenaries-who-helped-turn-tide-on-boko-haram

Sofaer, Abraham D. "Responses to Terrorism: Targeted Killing Is a Necessary Option." *San Francisco Chronicle*, March 26, 2004, http://www.sfgate.com/opinion/openforum/article/Responses-to-Terrorism-Targeted-killing-is-a-2775845.php

Sofge, Erik. "The Inside Story of the SWORDS Armed Robot 'Pullout' in Iraq: Update." *Popular Mechanics*, October 30, 2009, http://www.popularmechanics.com/technology/gadgets/a2804/4258963/

Springer, Paul J. *Military Robots and Drones*. Santa Barbara, CA: ABC-CLIO, 2013.

SRI International. "The Centibots Project." http://www.ai.sri.com/centibots/

Stanton, Doug. *Horse Soldiers: The Extraordinary Story of a Band of U.S. Soldiers Who Rode to Victory in Afghanistan*. New York: Scribner, 2009.

Stegherr, Laura K. "UAV DET Launches Final Pioneer Flight." *Navy News Service*, November 8, 2007, http://www.navy.mil/submit/display.asp?story_id=32916

Steinhoff, U. "Killing Them Safely: Extreme Asymmetry and Its Discontents," in B. J. Strawser, ed., *Killing by Remote Control: The Ethics of an Unmanned Military*. Oxford: Oxford University Press, 2013.

Strassler, Robert B., and Richard Crawley. *The Landmark Thucydides: A Comprehensive Guide to the Peloponnesian War*. New York: Free Press, 2008.

Strauss, Barry. *Masters of Command: Alexander, Hannibal, Caesar, and the Genius of Leadership*. New York: Simon & Schuster, 2012.

Strawser, B. J. "Moral Predators: The Duty to Employ Uninhabited Aerial Vehicles." *Journal of Military Ethics* 9, no. 4 (2009): 342–368.

Strickland, Matthew, and Robert Hardy. *The Great Warbow: From Hastings to the Mary Rose*. Stroud, UK: Sutton, 2005.

Sullins, J. P. "RoboWarfare: Can Robots Be More Ethical Than Humans on the Battlefield?" *Ethics and Information Technology* 12, no. 3 (2010): 263–275.

Sundvall, Timothy J. "Robocraft: Engineering National Security with Unmanned Aerial Vehicles." Maxwell Air Force Base, AL: School of Advanced Air and Space Studies, 2006.

Swinson, Mark L. "Battlefield Robots for Army XXI." Carlisle Barracks, PA: U.S. Army War College, 1997.

"Taliban Confirm Commander's Death." *BBC News*. August 25, 2009, http://news.bbc.co.uk/1/hi/world/south_asia/8220762.stm

Taulbee, James L. "Soldiers of Fortune: A Legal Leash for the Dogs of War?" *Defense Analysts* 1, no. 3 (1985): 187–203.

Terkel, Amanda. "Veterans in New Congress Fewest since World War II." *Huffington Post*, November 17, 2012, https://www.huffingtonpost.com/2012/11/17/veterans-congress-fewest_n_2144852.html

Thompson, Mark. "The Curse of 'Friendly Fire.'" *Time*, June 10, 2014, http://www.time.com/2854306/the-curse-of-friendly-fire/

Thoreau, Henry David. *Civil Disobedience*. Bedford, MA: Applewood Books, 2000. Originally published 1849.

Tilghman, Andrew. "New Medal for Drone Pilots Outranks Bronze Star." *Military Times*, February 13, 2013.

"Toys against the People, or Remote Warfare." *Science for the People Magazine* 5 no. 1 (1973): 8–10, 37–42.

Tyson, Ann Scott, and Josh White. "Top Two Air Force Officials Ousted." *Washington Post*, June 6, 2008, http://www.washingtonpost.com/wp-dyn/content/article/2008/06/05/AR2008060501908.html

United Nations. "Biological Weapons Convention." 2009, http://www.unog.ch/80256EE600585943/(httpPages)/04FBBDD6315AC720C1257180004B1B2F?OpenDocument

United Nations. "Chemical Weapons Convention." 2009, http://www.onog.ch/80256EE600585943/(httpPages)/4F0DEF093B460B4C1257180004B1B30?OpenDocument

United Nations Charter. Chapter 6: Pacific Settlement of Disputes. Avalon Project. Yale Law School, Lillian Goldman Law Library, http://avalon.law.yale.edu/20th_century/unchart.asp

United States Air Force. "B-2 Spirit Fact Sheet." December 16, 2015, http://www.af.mil/About-Us/Fact-Sheets/Display/Article/104482/b-2-spirit/

United States Air Force. "MQ-1B Predator." September 23, 2015, http://www.af.mil/About-Us/Fact-Sheets/Display/Article/104690/mq-1b-predator/

United States Air Force. "MQ-9 Reaper." September 23, 2015, http://www.af.mil/About-Us/Fact-Sheets/Display/Article/104470/mq-9-reaper/

United States War Department. *Official Records of the War of the Rebellion*, 128 volumes. Washington, DC: Government Printing Office, 1880–1901.

Utterback, Mitch. "How ISIS Is Turning Commercial Drones into Weapons in the Battle for Mosul." *Fox News*. January 25, 2017, http://www.foxnews.com/tech/2017/01/25/how-isis-is-turning-commercial-drones-into-weapons-in-battle-for-mosul.html

Van Joolen, Vincent J. "Artificial Intelligence and Robotics on the Battlefields of 2020?" Carlisle Barracks, PA: U.S. Army War College, 2000.

Vinge, Vernor. "The Coming Technological Singularity: How to Survive in the Post-Human Era." NASA Vision-21 Symposium, Westlake, OH, March 30–31, 1993.

Vlahos, Kelley Beaucar. "Handful of Lawmakers Send Their Kids to War." *Fox News*, March 28, 2003, http://www.foxnews.com/story/2003/03/28/handful-lawmakers-send-their-kids-to-wa-896091790.html

Wagner, William. *Lightning Bugs and Other Reconnaissance Drones*. Fallbrook, CA: Aero Publishers, 1982.

Wagner, William, and William P. Sloan. *Fireflies and Other UAVs*. Leicester, UK: Midland, 1992.

Walzer, Michael. *Arguing about War*. New Haven, CT: Yale University Press, 2005.

Walzer, Michael. "Targeted Killing and Drone Warfare." *Dissent*, January 11, 2013, https://www.dissentmagazine.org/online_articles/targeted-killing-and-drone-warfare

"Warplanes: The Air Force Loses Control of the Lower Altitudes." *Strategy Page*. February 19, 2013, https://www.strategypage.com/htmw/htairfo/articles/20130219.aspx

Washington Headquarters Services, Directorate for Information Operations and Reports. "Selected Manpower Statistics." Washington, DC: Government Printing Office, 1997.

Werrell, Kenneth P. *The Evolution of the Cruise Missile*. Maxwell Air Force Base, AL: Air University Press, 1985.

West, Bing. *No True Glory: A Frontline Account of the Battle for Fallujah*. New York: Bantam Books, 2006.

Whitlock, Craig. "U.S. Airstrike That Killed American Teen in Yemen Raises Legal, Ethical Questions." *Washington Post*, October 22, 2011, https://www.washingtonpost.com/world/national-security/us-airstrike-that-killed-american-teen-in-yemen-raises-legal-ethical-questions/2011/10/20/gIQAdvUY7L_story.html?utm_term=.04ae4febb22c

Whittle, Richard. *Predator: The Secret Origins of the Drone Revolution*. New York: Henry Holt, 2014.

Williams, Brian Glyn. *Predators: The CIA's Drone War on Al Qaeda*. Washington, DC: Potomac Books, 2013.

Williams, Brian Glyn, Matthew Fricker, and Avery Plaw. "New Light on the Accuracy of the CIA's Predator Drone Campaign in Pakistan." *Terrorism Monitor* 8, no. 4 (November 11, 2010): 8–13.

Williams, Dan. "Israel Plans Laser Interceptor 'Iron Beam' for Short-Range Rockets." *Reuters*, January 19, 2014, https://www.reuters.com/article/us-arms-israel-interceptor-idUSBREA0I06M20140119

Wilson, Mark. "CIA on the Verge of Lawsuit." *Seer Press News*, August 5, 2010, www.seerpress.com/cia-on-the-verge-of-lawsuit/3341/

Wong, Wilson W. S. *Emerging Military Technologies: A Guide to the Issues*. Santa Barbara, CA: Praeger, 2013.

Woods, Chris. *Sudden Justice: America's Secret Drone Wars*. New York: Oxford University Press, 2015.

Woodward, Bob. *Bush at War*. New York: Simon & Schuster, 2002.

World Nuclear Association. "Uranium Enrichment." May 2017, http://www.world-nuclear.org/information-library/nuclear-fuel-cycle/conversion-enrichment-and-fabrication/uranium-enrichment.aspx

"Wounded Warrior Amputee Softball Team." 2012, http://www.woundedwarrioramputeesoftballteam.org/

Yenne, Bill. *Birds of Prey: Predators, Reapers and America's Newest UAVs in Combat*. North Branch, MN: Specialty Books, 2010.

"'Zarqawi' Beheaded U.S. Man in Iraq." *BBC News*, May 13, 2004, http://news.bbc.co.uk/1/hi/world/middle_east/3712421.stm

Zenko, Micah. "100% Right 0% of the Time: Why the U.S. Military Can't Predict the Next War." *Foreign Policy*, October 16, 2012. http://foreignpolicy.com/2012/10/16/100-right-0-of-the-time/

Index

Abu Ghraib prison, 113
Adams, Douglas, 198
Adams, John, 122
AeroVironment, 101
Air defense system, 9, 75, 180, 182, 183, 185, 186, 204, 213
Air War Plans Division 1 (AWPD-1), 160
Aldrin, Edwin "Buzz," 67
Alexander the Great, 33, 207–8
Al Qaeda, 12–13, 20, 85–86, 90, 108, 111, 127–28, 137–41; attacks by, 16; declarations of war by, 14–15; goals, 13, 18–19; violations of laws of armed conflict, 117. *See also* September 11 attacks
Al Qaeda in Iraq (AQI), 91, 105
Al Qaeda in the Arabian Peninsula (AQAP), 108–12
American Civil Liberties Union, 143
Analytical engine, 61
Antiballistic Missile Treaty (1972), 53
Antisatellite weapons, 76
Arab Spring (2012), 27
Archytus, 59
Armstrong, Neil, 67
Arnold, Benedict, 105–6
Arnold, Henry Hap, 160

Artificial intelligence: definition, 9–10; performance, 11; predictions, 11–12
Artillery, development, 40
Asimov, Isaac, 196–97
Assad, Bashar al-, 95–96
Assassination, 77, 126, 140, 166–67
Atlanta, Battle of (1864), 163–64
Atrocities, 95–96, 157, 161, 165–66, 203–4
Atta, Mohamed, 109
Authorization for the Use of Military Force (2001), 85–86, 112, 140–42
Automata, 60–61
Autonomy: advantages of, 151–52; classes of, 6–8; definition, 6–7; lethal decision-making and, 166–67, 193
Autopilot systems, 73–74
Awlaki, Abdulrahan al-, 110–13
Awlaki, Anwar al-, 108–12, 142
Awlaki, Nasser al-, 109–10

Babbage, Charles, 61
Balkan Wars (1990s), 71
Banna, Ibrahim al-, 111
Banu Musa brothers, 59
Barbary Wars, 122
Battlestar Galactica, 202
Berg, Nicholas, 104

BigDog, 100
Bin Laden, Osama, 13–14, 17–20, 86,
    88, 91, 110, 140; death of, 133
Biological weapons, 90
Blackwater USA, 104–5
Bonaparte, Napoleon, 26
Boole, George, 61
Boston Dynamics, 100
Boxer Rebellion, 122
Brodie, Bernard, 51
Brooks, Rodney, 68
Builder, Carl, 80–81
Bush, George W., 16–17, 20, 53, 77,
    85–86, 89–90, 92, 125, 128, 132,
    140–41, 217; messages to Congress,
    17–19

Calley, William, 161
Cameron, James, 97
Camp David Accords, 181
Čapek, Karel, 195–96
Carter, Jimmy, 126
Centibots, 202–3
Central Intelligence Agency (CIA), 14,
    16–17, 77, 84, 109, 117–18, 127, 138,
    140, 143
Chalker, Jack L., 198–200
Challenger Deep, 97
Chechen War, 172
Chemical weapons, 90
Chiarelli, Peter, 133
Child soldiers, 96, 165
China, 135–36
Chinese Civil War, 48
Churchill, Winston S., 45
Cincinnatus, Lucius Quintus, 120
Clarke, Richard A., 16
Clausewitz, Carl von, 129, 153–54
Claymore antipersonnel mine, 93
Close air support, 88–89, 101–2
Cognition, 10–11
Cold War, 52, 69–70, 74, 99, 103–4
Colonization, 28, 43–44
Conscription, 206
Coolidge, Calvin, 122
Cortes, Hernan, 43
Counterterrorism, 3

Crossbow, 35–36
Cruise missile, definition, 6
Crusades, 35
Cuban Missile Crisis (1962), 49
Curiosity (spacecraft), 68
Cyberattack, 8, 186–92
Cybercrime, 170–71
Cyber espionage, 170, 177–79
Cyber security, 179
Cyber warfare, 125, 169–70; Chinese,
    178–80; Israeli, 184; Russian,
    173–78
Cyborg, definition, 9

Da Vinci, Leonardo, 59–60
Davis, Jefferson, 130, 163
Deepsea Challenger, 97
Descartes, Rene, 10–11
Desertion, 107, 207–8
Deterrence, 51–53, 180; nuclear, 51
Devol, George, 69
Dewey, George, 156–57
Difference engine, 61
Douhet, Giulio, 53–54
Drone: cruise missile, 6; definition, 6;
    programming limits, 6
Drone strikes: accuracy, 139; collateral
    damage, 139
Drone warfare, media coverage, 1, 5
Dual-use technology, 214–15

Einstein, Albert, 45–46
Eisenhower, Dwight D., 52, 123
Espionage, 117
Ethics, 3; definition, 148; historical
    precedents, 149–50; of warfare, 138
Executive Order 11905, 126
Executive Order 12333, 126
Explosive ordnance disposal (EOD),
    92–94, 98
Exponent (corporation), 93

Fallujah, Battle of, 159
FBI, 13
Feudalism, 34–35
Firearms: development of, 37–42;
    production of, 42

Fire Fly (drone), 70
Fisher, Vilyam, 75
Flame (malware), 190–92
Ford, Gerald, 123, 126
Foreign Intelligence Surveillance Act,
    92, 142
Foster-Miller (corporation), 94
French Revolution (1789), 26
Future of Life Institute, 220
Future predictions, 195

Gagarin, Yuri, 67
Gates, Horatio, 120
Gates, Robert M., 80, 83–84, 195
General Atomics, 88–89
General Orders Number 100 (1863),
    115–16, 144
Geneva Convention Relative to
    Prisoners of War (1929), 116–17,
    154–55
Geneva Convention Relative to
    Prisoners of War (1949), 116,
    135, 139
Genocide. *See* Atrocities
Georbot, 175–76
Germanwings disaster (2015), 73–74
Gibbs, Robert, 112
Global Hawk (RQ-4), 5–6, 8, 80, 133
Global positioning system (GPS), 76,
    209–10
Goertz, Raymond, 69
Goliath, 64
Grant, Ulysses S., 163
Gray Eagle, 101–2
Greene, Harold, 82
Grotius, Hugo de, 152–53
Guantanamo Bay detention facility,
    113, 142–43
Gulf of Tonkin incident, 155
Gunpowder: advantages of, 43–44;
    disadvantages of, 42; as RMA,
    36–44

Hague Conventions (1899 and 1907),
    154, 215
Hamas, 184; violations of laws of
    armed conflict, 117

Hamilton, Alexander, 120
Harding, Warren G., 122
Hawking, Stephen, 220
Hayes, Rutherford B., 131
Hellfire missile (AGM-114), 88–89,
    91, 110
Heydrich, Reinhard, 166
Hezbollah, 184, 192
Hobbes, Thomas, 10–11
Holocaust, 96, 166
Homer, 58
Hood, John Bell, 163
Hull, William, 130
Human behavioral limits, 7
Human–robot interactions, 101
Hume, David, 10–11
Hundred Years' War, 34, 36
Hussein, Saddam, 90–91, 159, 182
Hydrogen bomb, 50

Improvised explosive devices (IEDs),
    83, 90–92, 94
Industrial Revolution, 26
Innocent II, 34–35, 214
Innovation, as a continual process, 25
Inspire, 111
Intelligence, surveillance, and
    reconnaissance (ISR), 84, 86–88, 91,
    101, 132
Intercontinental ballistic missile
    (ICBM), 49–50
International Atomic Energy Agency
    (IAEA), 47, 184
International Committee of the Red
    Cross, 116
International Criminal Court, 214
Internet, creation of, 27
Iran-Iraq War, 182–83
iRobot (corporation), 93–94
Iron Beam, 185
Iron Dome, 185
Islamic State, 113, 203–5
Israel, regional conflict and, 181–82, 184
Israel Defense Force (IDF), 180

Jackson, Andrew, 119
Jacquard, Joseph Marie Charles, 61

Jacquet-Droz, Pierre, 60
Jazari, Abu al-, 59
Jefferson, Thomas, 122
Johnson, Lyndon B., 123, 155, 161
Johnston, Joseph E., 163
Just War, 152–53, 162, 167

Kahn, Herman, 51
Kaiko, 97
Karensky, Alexander, 26
Keene, John M., 93
Kennedy, John F., 65, 123, 161
Kennedy, Joseph P., Jr., 65
Kettering, Charles, 63
Kettering Bug, 63–64
Khan, Samir, 111–12, 142
Killing, act of, 82–83, 85
Kim, Jong-Un, 217
Knauss, Friederich von, 60
Knights, 34, 36–37
Korean War, 122, 134
Kurzweil, Ray, 11–12
Kyle, Chris, 83

Lateran Council (1139), 34–35
Laws of armed conflict, 101–2; ethical
    aspects, 148
Lee, Barbara, 86
Lee, Robert E., 130, 163
Lehel, Marcel "Guccifer," 177–78
LeMay, Curtis, 161
Lieber, Francis, 115–16, 144
Lightning Bug (drone), 70
Limbaugh, Rush, 9
Lincoln, Abraham, 99, 119–22, 130–31,
    137, 163
Lindh, John Walker, 107–8
Low Cost Autonomous Attack System
    (LOCAAS), 212–13
Lubitz, Andreas, 73–74
Luna 9, 67
Lunokhod I, 67

MacArthur, Douglas, 132
Madison, James, 121, 129–30
Maillardet, Henry, 60
Manhattan Project, 44–47, 121

Manila Bay, Battle of (1898), 156–57
Marathon, Battle of, 32–33
MARCbot, 93
Marescaux, Jacques, 72
Marianas Trench, 97
Mariner 10, 67
Marshall, John, 119
Martin, Matt, 82–83
Maude, Timothy, 82
McClellan, George, 131
McKinley, William, 122, 131, 156
McNamara, Robert S., 161
Meade, George, 163
Medieval period, 34
Mehsud, Baitalluh, 1
Meir, Golda, 182
Melian dialogue, 161–62
Mercenaries, 102–5, 138–39, 205
Mexican War, 130, 137
Microsoft Corporation, 8
Military discrimination, definition,
    158
Military engineering, 33
Military necessity, definition, 162–63
Military robot: accidents, 213;
    advantages of, 21, 206–8, 211;
    bans, 210–11, 214–15, 217; cost, 94,
    99, 101, 205; dangers, 20, 218–20;
    definition, 8; development, 79–80,
    212–14, 220; disadvantages, 219–20;
    discrimination, 162, 211; effects
    upon human morality, 168; Israel,
    185; legal aspects, 142; lethality
    of, 95–96; military necessity, 166;
    numbers, 92–93, 99, 101, 206;
    production, 99; programming and
    logic, 166; proportionality, 157–58;
    Republic of Korea, 193; as RMA, 55;
    revolt, 210; Russia, 171–72
Mine-resistant, ambush protected
    (MRAP) vehicles, 83
Minsky, Marvin, 11
Mosely, Michael, 84
Musk, Elon, 220
Mutually assured destruction
    (MAD), 52
My Lai Massacre, 161

Neumann, John von, 66
Nixon, Richard M., 123–24, 136, 161
No-fly zones, 156
North Atlantic Treaty Organization
    (NATO), 134, 172–75
Northern Alliance (Afghanistan), 87
Nuclear deterrence, 51
Nuclear escalation, 51–52, 54
Nuclear Nonproliferation Treaty, 210,
    214, 216
Nuclear weapons: development,
    45–46, 49–51; France, 48–49; India,
    48; Iran, 53, 185–91; Iraq, 183; Israel,
    48; Pakistan, 48, 218–19; People's
    Democratic Republic of Korea,
    48, 53, 192–93, 216–17; People's
    Republic of China, 48–49; as RMA,
    44–53; Soviet Union, 48, 50–51;
    Syria, 183–84; United Kingdom,
    48–49; usage in warfare, 47, 49–50

Obama, Barack, 1, 92, 111, 125–26, 133,
    142–43
Ogarkov, Nikolai Vasilyevich, 24
Operation Allied Force, 211–12
Operation Desert Storm. *See* Persian
    Gulf War
Operation Enduring Freedom, 86,
    93–94, 128, 132
Operation Iraqi Freedom, 71–72, 89,
    91, 94, 104, 128, 159
Operation Linebacker II, 113, 123
Operation Paperclip, 66
Operation Rolling Thunder, 155
Opportunity (spacecraft), 67–68

PackBot, 93–94
Pakistan, 1, 88
Panetta, Leon, 109
Pape, Robert, 53–54
Pathfinder, 67
Patriot missile battery, 9, 11
Peloponnesian War, 31–32
People's Republic of China, 135–36
Perry, William J., 14
Persian Gulf War, 71–72, 133, 156, 158–59
Phalanx, as RMA, 30–33

Philippine War, 122
Phoenix Close-In Weapons System
    (CIWS), 213
Piccard, Jacques, 97
Pioneer (RQ-2), 71–72
Pioneer 10, 68
Planetbot, 69
Polk, James K., 130
Powers, Francis Gary, 74–75
Precision-guided munitions (PGMs),
    158–59
Predator (RQ-1/MQ-1), 8, 72, 77, 80,
    84, 87–89, 91, 101–2, 110–11
Project Aphrodite, 65–66
Proportionality, definition, 148–49
Pryor, Frederic, 75
Punic Wars, 165
Punishment strategies, 53–54
Putin, Vladimir, 176, 178

QinetiQ, 94
Quasi-War with France, 122
Quesada, Elwood Pete, 88

"Rape of Nanking," 157
Raven (RQ-11), 101
Reagan, Ronald, 53, 126
Reaper (RQ-1/MQ-1), 8, 80, 89
Remotely-piloted vehicle (RPV),
    definition, 5–6
Rensselaer, Stephen van, 130
Revolution, definition, 25–26
Revolutionary War (U.S.), 129
Revolutionary warfare, definition,
    26, 129
Revolution in military affairs
    (RMA): advantages, 25, 27, 32,
    36–38; causes, 28–29; definition, 24;
    expansion, 28; timing, 27, 40
Robosurgery, 72–73
Robot: advantages over humans, 10;
    control issues, 98; definition, 5;
    early visions of, 57–61; industrial
    applications, 69; medical, 72, 101;
    programming and logic, 5, 8, 147,
    149–50, 168; sensory capabilities, 5.
    See also Military robot

Robotic system, definition, 5
Roman Empire, 33, 118, 120, 165
Roosevelt, Franklin Delano, 45–46,
    119–22, 131, 137
Roosevelt, Theodore, 122
Rules of Land Warfare, The, 116
Rumsfeld, Donald, 90
Russia-Estonia Conflict (2007), 173–75
Russia-Georgia Conflict (2008), 175–76
Russian Business Network (RBN), 171
Russian Revolution (1917), 26
Russia-Ukraine conflict, 1–2, 176–77

SA-2 missile, 75
Saberhagen, Fred, 197
Sadat, Anwar, 181
Sanchez, Robert M., 74
San Patricio Battalion, 107
Santiago, Battle of (1898), 156
Sattler, John, 159
Scheinman, Victor, 69
Scott, Winfield, 107, 119, 130
Sentinel (RQ-170), 75–76, 133
September 11 attacks, 12, 16, 82,
    127, 134
Shamoon virus, 192
Sherman, William T., 163–64
Shibh, Ramzi al-, 109
Shinseki, Eric, 90, 133
Singer, P. W., 96
Sojourner, 67
Somalia, 71
Space race, 66–67
Spanish-American War, 156
Sperry, Elmer, 63
Spirit (spacecraft), 67–68
Sputnik, 66, 76
Stalin, Josef, 47–48
Stanford Arm, 69–70
Star Trek, 200–201
Star Wars, 201
Stuxnet, 187–92
Submarine, 97–98
Suicide bombing, 1
Suicide drones, 1–2
Sun Tzu, 151–52
Surveyor I, 67

Swarming, 68, 202–3
Syrian Civil War, 95–96
Szilárd, Léo, 45–46

Taiwan, 180
Taliban, 17, 19, 82, 87, 86–89, 107
Talon Special Weapons Observation
    Reconnaissance Detection System
    (SWORDS), 94–95, 98
Targeted killing program, 77, 91–92,
    109–12, 127, 140, 143. *See also*
    Assassination
Taylor, Zachary, 130
Technology, first adopters, 23–24
Tehrik-e-Taliban Pakistan (TTP), 1
Teledyne-Ryan, 70
Telepresence, 72–73
Telesurgery, 72
Teletank, 64
Tenet, George, 77
Terminator, The, 201–2
Terrorism, 12, 14; definition, 136–37
Tesla, Nikola, 61–63
Thermite, 1–2
Thirty Years' War, 152
Thucydides, 161–62
Thunderbolt (A-10), 88, 102
Tonkin Gulf Resolution, 155
Treason, 105–8, 208
Trenchard, Hugh, 53–54
Trieste, 97
Truman, Harry S., 52, 122, 131–32
Trump, Donald, 125–26
Tsar Bomba, 50

U-2 (airplane), 74
Ukraine, 1–2
Uniform Code of Military Justice,
    117
Unimate robot, 69
United Nations Charter, 124–25, 134
United Nations Security Council, 125,
    134–36, 141, 143, 210
Unmanned ground vehicle (UGV),
    92–94, 100, 205
Unmanned underwater vehicle
    (UUV), 96–98

U.S. Air Force, 77, 81–82, 84–85, 101–2
U.S. Army, 80–81, 92, 95, 99, 101–2
U.S. Bill of Rights, 106–8, 114
U.S. Civil War, 99, 121, 130, 137,
    163, 208
U.S. Constitution, 106–8, 114, 118–22,
    134, 137
U.S. Department of Defense, 127–28
U.S. Department of State, 117, 128
U.S. military: interservice rivalry,
    101–2; preferences for combat, 100;
    prepositioned equipment, 100
U.S. Navy, 61–62, 81–82, 84–85,
    96–97, 180
U.S. presidency, wartime powers of,
    122–24, 126, 131, 138, 141
USS Cole bombing, 16
U.S. War on Terror, 1–4, 12–15, 20,
    84–85, 107–8, 112–13, 127–28,
    134–35, 139–42, 144, 162, 167;
    effect upon alliances, 18–19
Uzbekistan, 87

V-1 flying bomb, 64–65
V-2 missile, 66
Vaucanson, Jacques de, 60
Venera 8, 67
Vietnam War, 70, 113, 123, 136, 132,
    155, 161
Viking I, 67

Voyager I, 68
Voyager II, 68

Walker, Walton, 82
Walsh, Don, 97
War of 1812, 120–21, 129–30
War of the Triple Alliance, 164–65
War on Terror. See U.S. War on
    Terror
War Powers Resolution, 123–26
Washington, George, 120, 129
Weaponry: bans upon, 34–36; melee, 33,
    34; non-lethal, 167; primitive, 29–30
Weapons of mass destruction, 90, 217
Wiley, Wilford J., 65
Wilson, Woodrow, 122, 131, 137
Winter War (1939–1940), 64
Worcester v. Georgia, 119
World Trade Center, 93
World Trade Center attack (1993), 12–13
World War I, 44–45, 62–63, 103
World War II, 121; European Theater,
    159–60; Pacific Theater, 47, 157
Wright, Orville and Wilbur, 62
Wynne, Michael W., 84

Yemen, 108, 110–12
YouTube, 109–10

Zarqawi, Musab al-, 91, 104–5

About the Author

PAUL J. SPRINGER received his doctorate in history from Texas A&M University and has taught at that institution, the U.S. Military Academy at West Point, and the Air Command and Staff College (ACSC). He is currently a professor and the chair of the Department of Research at ACSC. Dr. Springer is a Senior Fellow of the Foreign Policy Research Institute and the editor of the *History of Military Aviation* and *Transforming War* series with the Naval Institute Press. His published works include *America's Captives: Treatment of POWs from the Revolutionary War to the War on Terror*; *Military Robots and Drones: A Reference Handbook*; *Transforming Civil War Prisons: Lincoln, Lieber, and the Politics of Captivity*; *Cyber War: A Reference Handbook*; and *9/11 and the War on Terror: A Documentary and Reference Guide*. Dr. Springer also served as editor of the *Encyclopedia of Cyber Warfare*.